14. Growth, Decay and Oscillation 1

THE OPEN UNIVERSITY
Technology/Mathematics: A Second Level Course

**MODELLING
BY MATHEMATICS**
TM281

Block 5
EVOLUTION

14. Growth, Decay and Oscillation 1
15. Growth, Decay and Oscillation 2
16. Review

Prepared by the Course Team

THE OPEN UNIVERSITY PRESS

THE MODELLING BY MATHEMATICS COURSE TEAM

David Blackburn   (Chairman)

Phil Ashby   (BBC)
Keith Attenborough   (Technology)
Gerald Copp   (Editor)
Peter Cox   (Student Computing Service)
Bob Davies   (Senior Counsellor)
Judy Ekins   (Mathematics)
Andy Harding   (Course Manager)
Roger Harrison   (Institute of Educational Technology)
Don Hurtley   (Staff Tutor)
Maurice Inman   (Staff Tutor)
David Johnson   (Student Computing Service)
Roy Knight   (Mathematics)
Ernest Law   (Staff Tutor)
Owen Lawrence   (Staff Tutor)
Daniel Lunn   (Mathematics)
Patricia McCurry   (BBC)
Alistair Morgan   (Institute of Educational Technology)
Colin Robinson   (BBC)
John Sparkes   (Technology)
Robert Tunnicliffe   (Mathematics)
Mirabelle Walker   (Technology)
Geoffrey Wexler   (Technology)

The Open University Press
Walton Hall, Milton Keynes
MK7 6AA

First published 1977. Revised edition 1979. Reprinted 1982.

Designed by the Media Development Group of the Open University.

Produced in Great Britain by
Technical Filmsetters Europe Limited, 76 Great Bridgewater Street, Manchester M1 5JY

ISBN 0 335 06297 0

This text forms part of an Open University course. The complete list of units in the course appears at the end of this text.

For general availability of supporting material referred to in this text, please write to: Open University Educational Enterprises Limited, 12 Cofferidge Close, Stony Stratford, Milton Keynes, MK11 1BY, Great Britain.

Further information on Open University courses may be obtained from the Admissions Office, The Open University, P.O. Box 48, Walton Hall, Milton Keynes, MK7 6AB.

2.2

CONTENTS

AIMS

The aims of this unit are:

1 To show you how second-order differential equations can be set up to represent various physical processes.

2 To show you how to find the general solutions of some types of second-order differential equations.

OBJECTIVES

After reading this unit you should be able to

1 Distinguish between true and false statements concerning, or explain in your own words, the following terms:

> boundary conditions
> initial conditions
> sinusoidal

2 Find the general solutions of differential equations of the following kinds

(a) $\dfrac{d^2 y}{dx^2} = f(x)$ (SAQs 2, 5, 6, 7, 9 and 10)

(b) $\dfrac{d^2 y}{dx^2} - \lambda^2 y = 0$ (SAQs 16, 17, 20 and 26)

(c) $\dfrac{d^2 y}{dx^2} + \lambda^2 y = 0$ (SAQs 19 and 20)

(d) $a\dfrac{d^2 y}{dx^2} + b\dfrac{dy}{dx} + cy = 0$ $(a \neq 0)$ (SAQs 22, 24 and 25)

3 Identify a general solution of one of these equations and distinguish it from a particular solution. (SAQs 12 and 13)

4 Use a pair of initial or boundary conditions to obtain simultaneous equations for the two arbitrary constants occurring in the general solution of a second-order differential equation. Solve the simultaneous equations and hence find the particular solution. (SAQs 4, 5, 7, 8, 10, 11, 17, 19 and 20)

5 Verify that given expressions are solutions of particular differential equations. (SAQs 1, 14, 15, 18, 21 and 23)

6 Write down a differential equation to describe a particular model. Use the results of a model to make predictions about the behaviour of the system being modelled. (SAQs 2, 3, 6, 27 and 28)

STUDY GUIDE

Units 14 and 15 are an introduction to the topic of second-order differential equations. Together they form one complete package, but to help you to plan your study of them, the package has been split into two approximately equal units. You should remember, however, that Unit 14 is rather longer than Unit 15 and allocate your study time accordingly.

This week's work consists of studying Unit 14, *Growth, Decay and Oscillation 1*, listening to Sides 1 and 2 of Disc 8 and answering the assignment material.

The unit uses material discussed earlier in the course, such as the theory of falling bodies in Unit 6 and the modelling of heat flow in Unit 13. You will, of course, have to use much of the calculus introduced from Unit 7 onwards. Of particular importance is the discussion in Units 8 and 11 of indefinite integrals and arbitrary constants or constants of integration.

Section 4 of the unit starts with the derivation of a second-order differential equation for modelling the heat removed by a cooling fin. You will not be expected to reproduce the arguments included there, although you should be able to solve the resulting differential equation.

You should listen to Disc 8 at the point indicated in Section 3.3.

You can check on what you are expected to know by consulting the summaries and the objectives.

1 INTRODUCTION

In this unit and the next you will study second-order differential equations. You may remember that second-order differential equations were mentioned very briefly in Unit 8 when the idea of a differential equation was first introduced.

> What is the essential difference between a first-order differential equation and a second-order differential equation?

A first-order differential equation contains no derivatives other than first derivatives. A second-order differential equation includes at least one second derivative; it may also include first derivatives but will not include any derivatives of higher order than second. The following are examples of second-order differential equations.

$$\frac{d^2 y}{dt^2} = a \sin \omega t + by$$

$$a\frac{d^2 y}{dt^2} + b\frac{dy}{dt} + cy = 0 \quad (a \neq 0)$$

$$\frac{d^2 x}{dt^2} - p\frac{dx}{dt} = q \exp kt$$

$$\frac{d^2 y}{dx^2} = 0$$

$$\frac{d^2 z}{dx^2} - \lambda^2 z = 0$$

These examples all contain second derivatives and some also include first derivatives. None includes derivatives beyond the second. Differential equations with derivatives beyond the second are called third, fourth, fifth, etc. order differential equations, depending on whether the highest derivative appearing is third, fourth, fifth, etc. They are less often used in modelling than first or second-order differential equations and are not discussed in this course.

In Unit 8 and Block 4 you saw that a solution of a first-order differential equation is a mathematical equation relating one variable to another. It may, for example, be an equation giving population as a function of time. The solution of a second-order differential equation is also an equation relating one variable to another. Therefore, the solutions of the first two differential equations in my list above are of the form $y =$ some function of t; the third is of the form $x =$ some function of t.

> What is the form of the solution of the fourth and fifth differential equations?

The fourth is $y =$ some function of x; the fifth is $z =$ some function of x.

Units 14 and 15 will use models which give rise to two different sorts of second-order differential equations. The first sort of differential equation you will meet has the form

$$\frac{d^2y}{dx^2} = f(x)$$

How do you think this type of differential equation might be solved?

This second-order differential equation is similar to first-order differential equations of the type

$$\frac{dy}{dx} = f(x)$$

which you solved in Block 4. Perhaps this similarity has suggested to you that integration of $f(x)$ might solve such a second-order differential equation. In fact, this is the method of solution, although $f(x)$ must be integrated *twice*. You will see how this works in Section 2 of this unit.

The second sort of differential equation you will meet is the family of differential equations

$$a\frac{d^2y}{dx^2} + b\frac{dy}{dx} + cy = 0 \quad (a \neq 0)$$

where the constant parameters a, b and c may be positive or negative and b or c may be zero.

First you will learn how to solve such differential equations and then you will see how such equations arise from models of heat flow down a fin, of the oscillations of an object hanging on the end of a spring and of the oscillations of a pendulum. One of these models is in the last section of Unit 14, the rest are in Unit 15.

A few final words before you embark on learning how to solve second-order differential equations: As with first-order differential equations, your task is to find an equation which relates your variables. Once you have learned to use the rules which enable you to do this, solving second-order differential equations need be no harder than solving first-order ones. Being able to solve second-order differential equations will be an invaluable skill if you plan to study other courses which use mathematical modelling, because such modelling frequently involves second-order differential equations. Therefore, these two units are an important part of the course.

2 DIFFERENTIAL EQUATIONS OF THE FORM $d^2y/dt^2 = f(t)$

I shall start this section by looking at a very simple example of the differential equation

$$\frac{d^2y}{dt^2} = f(t)$$

This example will be the one where $f(t)$ is a constant:

$$\frac{d^2y}{dt^2} = a$$

where a is a constant.

This differential equation occurs in models which represent bodies falling under constant acceleration due to gravity because, as you should remember from Unit 7, if y is a displacement then d^2y/dt^2 is an acceleration.

When you have learned how to solve this differential equation it is only a small step to learn how to solve

$$\frac{d^2y}{dt^2} = f(t)$$

an equation which may model the motion of objects whose acceleration is *not* constant but varies with time.

2.1 A second-order differential equation for falling bodies

You carried out various calculations on bodies moving under the influence of gravity in Unit 6 and again in Units 10 and 11. I now want to represent the same motion using second-order differential equations since this approach can be used for the more general cases of varying acceleration. I shall start by showing you how such a second-order equation can be obtained and then go on to solve it.

Consider an object moving under the influence of gravity. Let its instantaneous height above the ground be Y metres, where distance upwards is regarded as positive.

As in previous units, I shall neglect air resistance. The body therefore falls with a constant acceleration which, in Britain, is very nearly a constant with value $9.8 \, \text{m s}^{-2}$ (to two significant figures).

How can acceleration be written in terms of Y?

It is represented by

$$\frac{d^2Y}{dT^2} \, \text{m s}^{-2}$$

where T is the time in seconds.

Y is considered to be positive upwards here, so d^2Y/dT^2 represents the *upwards* acceleration of the object. But we already know that the object is falling with a downward acceleration of $9.8 \, \text{m s}^{-2}$. Therefore

$$\frac{d^2Y}{dT^2} = -9.8 \tag{1}$$

This equation simply means that $-d^2Y/dT^2$ is equal to the acceleration due to gravity and, as you know, an equation which contains a second derivative is called a second-order differential equation. In equation (1) the time, T seconds, is the independent variable. The variable Y is the dependent variable.

The purpose of expressing 'falling under gravity' as a second-order differential equation is to show, through a well-known problem, how to set about solving such equations. I shall then be able to apply the method to cases where acceleration varies with time.

2.2 The general solution of the second-order differential equation for a falling body

Consider a situation in which an object is released from rest at a height of 100 m and falls freely. One of the equations in Unit 6 ($s = ut + \frac{1}{2}at^2$) tells you that it falls a distance $\frac{1}{2} \times 9.8T^2$ metres in T seconds (neglecting air resistance). Its height, Y metres, T seconds after release is therefore

$$Y = 100 - \frac{1}{2} \times 9.8T^2$$

$$= 100 - 4.9T^2 \tag{2}$$

With this result, you can work out the height of the object at all times between its release and its impact with the ground. Because equation (1) also describes the falling object, equation (2) should be a solution; that is, the function of T which equation (2) specifies should, if differentiated twice, given equation (1).

SAQ 1

By differentiating equation (2), verify that it is a solution of differential equation (1).

Equation (2) is one particular solution of differential equation (1).

Why is it a particular solution rather than a general one?

It is a particular solution because it is the solution for an object released from *rest* at a *height of 100 m*. A general solution should refer to *any* object, released at *any speed* at *any height*. The general solution should, therefore be an expression for Y in terms of T which

1 does not contain any derivatives (even first-order ones);

2 has a second derivative which satisfies differential equation (1) for all values of Y and T;

3 does not embody any specific information about when or where the object was released.

The equation you were given, equation (2), was one particular solution of equation (1) and I asked you to check that it was correct. Now I want to show you how to find the solution of the differential equation directly, step by step, without the aid of an algebraic solution, even though for all cases of constant acceleration the equation of Unit 6 can be used. Thus I am using the case of constant acceleration to demonstrate a more general method of solution.

The task may be stated as follows: 'given that the second derivative of a function is a constant, find the general form of that function'.

It will help you to remember that d^2Y/dT^2 may also be written

$$\frac{d}{dT}\left(\frac{dY}{dT}\right)$$

12

In words, the second derivative is just the derivative of the first derivative. So, if I integrate the expression for

$$\frac{d}{dT}\left(\frac{dY}{dT}\right)$$

I should expect to obtain an expression for dY/dT. This is a first-order differential equation and, as you know, integration again will give an expression for Y.

Thus, starting from equation (1)

$$\frac{d^2Y}{dT^2} = \frac{d}{dT}\left(\frac{dY}{dT}\right) = -9.8$$

If I integrate both sides and add an arbitrary constant of integration B, I get

$$\frac{dY}{dT} = \int(-9.8)\,dT$$

$$= -9.8T + B$$

You can check that this satisfies differential equation (1) by differentiating it, when you will recover the second-order equation.

> Have I solved the second-order differential equation?

Although this formula satisfies the second-order equation, it still contains a derivative and is therefore not yet a solution.

To complete the process I have to integrate both sides a second time, once again adding a constant of integration, which I shall call C.

$$Y = \int(-9.8T + B)\,dT$$

or

$$Y = -\tfrac{1}{2} \times 9.8T^2 + BT + C$$

$$= -4.9T^2 + BT + C \qquad (3)$$

Equation (3) is called the general solution of equation (1). It is a general solution because it embodies no specific information about where the object was at a particular time. It is possible to write down solutions specific to particular cases by giving values to B and C. B and C are referred to as the *arbitrary constants* in the solution. Notice that there was *one* arbitrary constant in a *first*-order differential equation and there are *two* in this *second*-order differential equation. You will find this to be true in general— a second-order differential equation has two arbitrary constants in its general solution.

Equation (3) stands for a whole family of solutions with different values of B and C.

> Equation (2) is a special case of equation (3); what are the values of B and C?

$B = 0$

$C = 100$

SAQ 2

SAQ 2

The acceleration due to gravity near the surface of the Moon is about one-sixth of the value near sea-level on the Earth. Write down the second-order differential equation describing a body falling towards the Moon at a point near the Moon's surface and find its general solution.

2.3 Particular solutions: initial and boundary conditions

Equation (3) contains two constants, *B* and *C*. As mentioned above, differential equation (1) is satisfied whatever the values of *B* and *C*. Solutions with specific values of *B* and *C* are called particular solutions. But how can a particular solution be found from the general solution; that is, how can values for *B* and *C* be found for a specific case? To answer this, let me remind you first about finding a value for the arbitrary constant in first-order differential equations.

In first-order differential equations, how did you give a value to the arbitrary constant in the general solution?

By using an initial condition; that is, a value for the dependent variable at a particular value of the independent variable.

In the first-order differential equations there was one arbitrary constant, and one specific piece of information was needed. In second-order differential equations there are *two* arbitrary constants, and, therefore, *two* specific pieces of information are needed.

One way in which this information can be given is to quote the value both of the dependent variable and of its first derivative at one particular value of the independent variable. Such information represents two *initial conditions*. In the example of a falling body, *Y* and dY/dT (that is, the height and the velocity) would be given at one particular value of *T*; probably, but not necessarily, $T = 0$. I can show how this works by modelling the motion of a pilot making an emergency jump out of an aeroplane.

initial conditions

What is the lowest height from which he can safely make such a jump? If the plane is too low, he might hit the ground before his parachute can open, and in that case he would probably be safer making an emergency landing in the plane.

I shall suppose that in the interval before his parachute opens I can model the pilot's fall by using differential equation (1) for freely falling bodies and its solution, equation (3).

It is convenient to measure time from the moment the pilot jumps. As before, I shall take height above ground level to be *Y* metres at time *T* seconds since the jump. I shall look only at vertical motion.

If he jumps from a height of 500 metres, $Y = 500$ when $T = 0$. Substituting these values for *Y* and *T* in equation (3)

$$Y = -4.9T^2 + BT + C$$

gives

$$500 = 0 + 0 + C$$
$$= C$$

This means that the second constant *C* occurring in the general solution is determined by the initial height of the pilot.

The constant *B* is determined by the initial velocity of descent. As a first model I shall suppose that the pilot's initial vertical velocity is zero. So $dY/dT = 0$ at $T = 0$. To apply this to equation (3) I first have to find dY/dT by differentiation.

$$\frac{dY}{dT} = -4.9 \times 2T + B \qquad (4)$$

But at $T = 0$, $dY/dT = 0$, so

$$B = 0$$

The equation modelling the pilot's motion is therefore

$$Y = -4.9T^2 + 500 \qquad (5)$$

This is the solution for this particular case. The arbitrary constants B and C have been replaced by the specific constants 0 and 500.

Suppose that the time between the pilot's jump and the opening of his parachute is five seconds. I can substitute $T = 5$ into equation (5) giving

$$Y = -4.9 \times 25 + 500$$

$$= 378 \quad \text{(to three significant figures)}$$

for the value of Y when the parachute opens.

What does this mean?

It means that he is still over 370 m above the ground when his parachute opens—which should give it sufficient time to slow his rate of descent.

What would have happened if he had jumped from 150 m and his parachute took six seconds to open?

In this case, equation (5) would have been

$$Y = -4.9T^2 + 150$$

since the initial height would have been 150 m instead of 500 m. Putting $T = 6$ into this equation gives

$$Y = 4.9 \times 36 + 150$$

$$= -26.4$$

for the value of Y when the parachute opens.

What does this mean?

A negative value of Y can have no meaning—the pilot would have hit the ground before his parachute opened!

SAQ 3 SAQ 3

A particular parachute can be fully open 4 seconds after the pilot leaves the plane. To be effective, it should be fully open before he is 200 m above the ground. What is the lowest height from which a pilot using this parachute could safely make an emergency jump?

In many cases, it is not appropriate to take the initial velocity to be zero. For example, the aeroplane might be gaining or losing height when the pilot jumps. The pilot would then have an initial vertical velocity imparted by the aeroplane. Consider a more general case in which the initial height is Y_0 m and the initial velocity is V_0 m s^{-1}, that is

$$Y = Y_0 \qquad \text{at} \quad T = 0$$

and

$$\frac{dY}{dT} = V_0 \qquad \text{at} \quad T = 0$$

To use these initial conditions I must let $T = 0$ in both the general solution, equation (3), and its derivative, equation (4). I then get two equations for the unknown constants B and C: from equation (3)

$$Y_0 = C$$

and from equation (4)

$$\frac{dY}{dT} = V_0 = B$$

Thus in this case B and C represent the initial upwards velocity and initial height.

Substituting Y_0 and V_0 into equation (3) gives

$$Y = -4.9T^2 + V_0T + Y_0$$

This is the solution for this more general case.

SAQ 4

SAQ 4

In SAQ 2 you found the general solution of the differential equation describing an object falling towards the Moon. Use the general solution to find the particular solution for an object dropped from a spaceship 500 m above the Moon's surface when the spaceship is travelling upwards at $1500\,\text{m s}^{-1}$.

SAQ 5

SAQ 5

Find the general solution of the differential equation

$$\frac{d^2X}{dT^2} = 5$$

Find the particular solution for the case $X = 3$, $dX/dT = 1$ when $T = 1$.

SAQ 6

SAQ 6

An international company is reviewing the relative growth of twenty factories which it owns. Each factory's output is one million units per year at the beginning of the survey (June 1976). It is found that in each case the growth can be modelled by supposing that the second derivative of the output with respect to time is a constant, equal to 0.005 million units per year^3.

(a) Obtain a common equation for the output (against time) for all twenty factories.

(b) One group of factories was found to grow more slowly than the others. How can this conclusion be reconciled with the supposition?

(c) If one of the factories starts off with a zero rate of growth, what would its subsequent output be according to the model?

Initial conditions refer to two pieces of information at the *same* value of the independent variable (time in the examples above). The other way in which information can be given is that two pieces of information are given at two *different* values of the independent variable. These two pieces of information might be the two values of the dependent variable, or the two values of its first derivative or one of each. Such information represents two *boundary conditions*.* In the particular case we have been examining, values of Y could be given at T_1 and T_2, or values of dY/dT at T_1 and T_2 or a value of Y at T_1 and a value of dY/dT at T_2.

boundary conditions

* *Two initial conditions will always lead to values for both the arbitrary constants, but sometimes even two boundary conditions do not enable both arbitrary constants to be found. In this course any boundary conditions given will be adequate for your needs— but remember that this is not always the case.*

Example

Suppose a stone is thrown upwards from the ground ($Y = 0$) and it reaches its maximum height after 3 seconds. What is the particular solution of differential equation (1) here? Use it to find how high the stone goes.

Since at its highest point the stone has zero velocity (that is $dY/dT = 0$), the boundary conditions are

$$Y = 0 \qquad \text{at} \quad T = 0$$

$$\frac{dY}{dT} = 0 \qquad \text{at} \quad T = 3$$

Substituting these values into the general solution, equation (3), and its first derivative, equation (4) gives

$$0 = C$$

and

$$0 = -3 \times 9.8 + B$$

Therefore $B = 29.4$.

The particular solution is

$$Y = -4.9T^2 + 29.4T$$

Since the stone reaches its maximum height after 3 seconds, the value of Y at $T = 3$ gives the maximum height.

$$Y_{max} = -9 \times 4.9 + 3 \times 29.4$$

$$= 44.1$$

The maximum height it reaches is 44.1 m.

SAQ 7

SAQ 7

Find the general solution of

$$\frac{d^2Y}{dT^2} = 1$$

Find the particular solution for the case where $Y = 3$ when $T = 0$ and $Y = 5$ when $T = 2$.

SAQ 8

SAQ 8

A stone was thrown downwards (with an unknown initial velocity) from a building fifty metres high.

An onlooker with a stop-watch found that it took T_0 seconds to reach the ground. Starting with the solution of the second-order differential equation for freely falling bodies, find how high the stone was T seconds after being thrown. That is, find an expression for Y as a function of T using the particular details of this case.

In summary, then, a second-order differential equation of the form $d^2Y/dT^2 = A$, where A is a constant, can be solved by integrating twice. This process introduces two arbitrary constants into the general solution. A particular solution can be found by using two initial or two boundary conditions to give values to the two arbitrary constants.

2.4 The solution of a differential equation of the form $d^2y/dt^2 = f(t)$

The same method of solution as the one summarized above can be used to find an equation for the distance when the acceleration is some known (non-constant) function of time.

Consider, for example, a car travelling along a straight road AB whose acceleration on one section of this road increases steadily from $2\,\mathrm{m\,s^{-2}}$ to $6\,\mathrm{m\,s^{-2}}$ over an eight-second period.

Because the acceleration increases 'steadily' a linear model is appropriate where a constant rate of increase of acceleration is supposed. Such a model describes the acceleration as varying with time T seconds according to the equation

$$\frac{d^2Y}{dT^2} = 2 + 0.5T \qquad (0 \leqslant T \leqslant 8) \tag{6}$$

where Y metres is the displacement of the car from the point A and $T = 0$ is the instant when the acceleration was $2\,\mathrm{m\,s^{-2}}$.

The solution of differential equation (6) is an expression showing how Y varies with T.

How can the general solution be found?

The differential equation can first be written in the form

As before, it is found by integrating differential equation (6) twice with respect to T.

$$\frac{d}{dT}\left(\frac{dY}{dT}\right) = 2 + 0.5T \qquad (0 \leqslant T \leqslant 8)$$

which means that

$$\frac{dY}{dT} = \int (2 + 0.5T)\,dT$$

Performing the integration gives

$$\frac{dY}{dT} = 2T + \frac{0.5T^2}{2} + B \qquad (0 \leqslant T \leqslant 8)$$

where B is an arbitrary constant.

A second integration gives

$$Y = \int \left(2T + \frac{0.5T^2}{2} + B\right)dT$$

$$= T^2 + \frac{0.5T^3}{6} + BT + C \qquad (0 \leqslant T \leqslant 8)$$

where C is a second arbitrary constant.

This is the general solution of differential equation (6).

SAQ 9

SAQ 9

Use the same method to find the general solution of the differential equations:

(a) $\dfrac{d^2Y}{dT^2} = -10 + 3T$

(b) $\dfrac{d^2Y}{dT^2} = \exp 2T + \exp T$

(c) $\dfrac{d^2y}{dt^2} = c\sin\omega t$, where c and ω are constants.

This method always works for a second-order differential equation of the form

$$\frac{d^2y}{dt^2} = f(t)$$

provided both $f(t)$ and the integral of $f(t)$ can be integrated.

Particular solutions can be found just as they were in Section 2.3—initial or boundary conditions are used. Try this for yourself in the next two SAQs.

SAQ 10

SAQ 10

(a) Find the particular solution of

$$\frac{d^2 Y}{dT^2} = 5T^2$$

given $Y = 3$, $dY/dT = 1$ at $T = 0$.

(b) Find the particular solution of

$$\frac{d^2 X}{dT^2} = 3 \cos T$$

given $X = 1$ at $T = 0$ and $X = 4$ at $T = \pi/2$.

SAQ 11

SAQ 11

The car whose motion was modelled by differential equation (6) is known to have been 10 m from A at $T = 0$ and 100 m from A at $T = 6$. Find the particular solution which describes how the displacement of this car, Y metres, varies with T.

How far did the car travel between $T = 0$ and $T = 3$?

2.5 Have I missed any of the solutions?

Since a differential equation has so many solutions, how can I ever be sure that the particular solution I am seeking can always be found from the general solution? The answer is that the term 'general solution' means one formula that includes *every* possible particular solution of the differential equation. Equation (3) thus models *every* falling body for which the suppositions of differential equation (1) are regarded as accurate enough. This is due to the presence of two arbitrary constants. This is generally true for all second-order equations: that is, the occurrence of two independent constants of integration in a solution of a second-order differential equation means that such a solution contains all possible particular solutions to it. Equally, the presence of one constant of integration in a solution to a first-order differential equation implies that there are no other solutions.

However, the statement that a solution containing two arbitrary constants includes all possible solutions to a second-order differential equation contains some pitfalls. For example, is

$$Y = -4.9T^2 + BT + DT$$

a general solution of equation (1), $d^2 Y/dT^2 = -9.8$?

It certainly contains two constants B and D; but it is not the general solution because I can set

$$B + D = E$$

where E is another constant. The equation may then be rewritten as

$$Y = -4.9T^2 + ET$$

which only contains one constant. It does not contain the second constant C occurring in the general solution (equation (3)).

Another point to notice is that

$$Y = -4.9T^2 + BT + DT + C$$

is *not* a new kind of general solution, different from equation (3); it is a solution, but it is not new because you can set $B + D = E$ again to reduce the equation to equation (3), the general solution obtained earlier.

SAQ 12

SAQ 12

Which of the following functions are, or can be reduced to, the general solutions of the differential equations immediately following them?

(a) $Y = T^5 + BT + C$

$$\frac{d^2 Y}{dT^2} = 20T^3$$

(b) $y = bt + \sin \omega t + ct + a$

$$\frac{d^2 y}{dt^2} = -\omega^2 \sin \omega t$$

(c) $y = bt + e^t + ct$

$$\frac{d^2 y}{dt^2} = e^t$$

In dealing with the problem of falling bodies I have considered numerical values of acceleration. That is, I have been concerned with the solutions of the differential equation (1), namely

$$d^2 Y/dT^2 = -9.8$$

If I use dimensioned quantities the arbitrary constants in the solution are not so obvious. For example, I can write

$$d^2 y/dt^2 = a$$

where a is the acceleration, supposed constant.

Integrating both sides twice with respect to t and adding a constant on each occasion gives the general solution

$$y = \frac{at^2}{2} + bt + c \tag{7}$$

where b and c are arbitrary constants of integration.

This formula contains five different letters each of which has a different role. When you see a solution like this you should always check that you know what each letter stands for and, in particular, which are variables, which parameters and which arbitrary constants.

SAQ 13

SAQ 13

State what each letter in equation (7) represents (independent variable, dependent variable, constant depending on initial position, constant depending on initial velocity, etc.).

2.6 Summary

If the second derivative of a variable y is a known function of the independent variable t, that is, if

$$\frac{d^2 y}{dt^2} = f(t)$$

the general solution can be found by performing two successive indefinite integrations: for example, if

$$\frac{d^2y}{dt^2} = t$$

$$\frac{dy}{dt} = \int t \, dt = \frac{t^2}{2} + b$$

$$y = \int \left(\frac{t^2}{2} + b \right) dt = \frac{t^3}{6} + bt + c$$

where b and c are arbitrary constants of integration.

When a second-order differential equation is used for modelling, the calculation of the particular solution requires the following steps:

(a)   Find the general solution.

(b)   Look for two additional pieces of information with which to calculate the arbitrary constants. These may be *initial conditions* of the form:

$$y = y_0 \qquad \text{at} \quad t = t_0$$

$$\frac{dy}{dt} = v_0 \qquad \text{at} \quad t = t_0$$

where t_0 is some value of t (often taken to be zero).

(c)   Substitute the initial conditions into the general solution or its derivative to obtain two equations for the arbitrary constants. Solve these equations for the arbitrary constants.

Sometimes the arbitrary constants can be calculated from *boundary conditions* which provide information about the unknown variable at two different values of the independent variable. For example, if

$$\frac{d^2y}{dt^2} = f(t)$$

the constants of integration can be calculated from the alternative pairs of boundary conditions

$$y = y_0 \qquad \text{at} \quad t = t_0$$

$$y = y_1 \qquad \text{at} \quad t = t_1$$

or

$$y = y_0 \qquad \text{at} \quad t = t_0$$

$$\frac{dy}{dt} = v_1 \qquad \text{at} \quad t = t_1$$

3 DIFFERENTIAL EQUATIONS OF THE FORM $a\,d^2y/dx^2 + b\,dy/dx + cy = 0$

In the introduction I mentioned that differential equations of the type

$$a\frac{d^2y}{dx^2} + b\frac{dy}{dx} + cy = 0 \qquad (a \neq 0)$$

were important, and for the rest of Unit 14 and all of Unit 15 you will be studying such differential equations and their solutions. This section is intended to teach you how to solve these differential equations; it does not attempt to show how these differential equations are used in modelling—that is done later after you have the necessary tools for solving the equations at your disposal.

When b and c are both zero then this differential equation reduces to

$$a\frac{d^2y}{dx^2} = 0$$

which is exactly the type of differential equation you have been solving in Section 2; the solution can be found by integrating twice. In all other cases, however, a different technique is needed to solve this type of differential equation.

The easiest case to look at first is when $b = 0$. The differential equation then has the form

$$a\frac{d^2y}{dx^2} + cy = 0$$

I can divide both sides by a and write $c/a = h$ to give

$$\frac{d^2y}{dx^2} + hy = 0 \tag{8}$$

In this equation, h may be either a positive or a negative constant. You will see shortly that the form of solution depends on whether h—that is, the coefficient of y—is positive or negative. I am therefore going to split equation (8) into its two possibilities (h positive or h negative) and discuss them separately.

When h is positive I can write $h = \lambda^2$. λ^2 must always be positive and so this is a way of indicating that h is positive. I can then write equation (8) as

$$\frac{d^2y}{dx^2} + \lambda^2 y = 0 \qquad (\lambda \neq 0) \tag{9}$$

and the sign of the coefficient of y is now unambiguous.

How do you think equation (8) can be written when the coefficient of y is negative?

As

$$\frac{d^2y}{dx^2} - \lambda^2 y = 0 \quad (\lambda \neq 0) \tag{10}$$

Now, if you met a *first*-order differential equation of the form $dy/dx + \lambda^2 y = 0$ or $dy/dx - \lambda^2 y = 0$ you would know that the solution was exponential. So let us try an exponential solution for both equation (9) and equation (10).

SAQ 14

Differentiate $y = p \exp kx$ twice (p and k are constants). Can it be the solution of either equation (9) or equation (10) under any circumstances?

SAQ 14 will have shown you that

$$y = p \exp kx$$

would be a solution to differential equation (9) if it were true that

$$k^2 = -\lambda^2$$

But squares of real numbers are always positive so this relationship cannot be true and $y = p \exp kx$ cannot be a solution of differential equation (9).

On the other hand, $y = p \exp kx$ can be a solution of differential equation (10), because for that it is necessary for

$$k^2 = \lambda^2$$

and this is quite possible; if $k = +\lambda$ or $-\lambda$ the relationship will hold.

So I have found a solution of differential equation (10) by a trial and error method. In the next section I will examine the solution of this differential equation further, returning to differential equation (9) in Section 3.2.

3.1   The solution of $d^2y/dx^2 - \lambda^2 y = 0$

You saw in SAQ 14 that

$$y = p \exp \lambda x$$

and

$$y = p \exp(-\lambda x)$$

are solutions of differential equation (10). But can either of them be the general solution? The answer is no, because neither contains *two* arbitrary constants. You will remember that in Section 2 we found two arbitrary constants occur in the general solution of a second-order differential equation. In the next SAQ you will try to find the general solution by differentiating several proposed 'solutions', each of which contains two arbitrary constants.

SAQ 15

Differentiate each of the following expressions for y twice and test if each is a solution of $d^2y/dx^2 - \lambda^2 y = 0$.

(a) $y = p \exp \lambda x + b$ $(p, b \neq 0)$

(b) $y = p \exp(-\lambda x) + cx$ $(p, c \neq 0)$

(c) $y = p \exp \lambda x + q \exp(-\lambda x)$ $(p, q \neq 0)$

You will have found that only item (c) in SAQ 15 satisfies the differential equation. There are many other proposed 'solutions' you could have tried, but item (c) above is the only one that satisfies the differential equation. It is therefore the general solution. That is,

$$y = p \exp \lambda x + q \exp(-\lambda x) \tag{11}$$

(where p and q are arbitrary constants) is the general solution of

$$\frac{d^2y}{dx^2} - \lambda^2 y = 0$$

As before, the key point in recognizing that this is the general solution of this second-order differential equation is the presence of two independent arbitrary constants. My assertion that I have the general solution here is backed up by a mathematical theorem—but that need not concern you!

Example

What is the general solution of the differential equation

$$\frac{d^2Y}{dX^2} - 25Y = 0?$$

Here 25 corresponds to λ^2. I can write the general solution, equation (11), as

$$Y = P\exp 5X + Q\exp(-5X)$$

where P and Q are arbitrary constants.

As you can see, once you know that the solution is a sum of two exponentials it is very easy to write down the general solution of any given differential equation.

SAQ 16

SAQ 16

Write down the general solution of each of the following differential equations

(a) $\dfrac{d^2Y}{dX^2} - 4Y = 0$

(b) $\dfrac{d^2Y}{dT^2} - 16Y = 0$

(c) $\dfrac{d^2X}{dT^2} - 9X = 0$

Particular solutions of this differential equation can be found from initial or boundary conditions, just as with the differential equations of Section 2. Because there are now values for two constants to be found, you will frequently find you have simultaneous equations to solve.

SAQ 17

SAQ 17

(a) Find the particular solution of

$$\frac{d^2Y}{dT^2} - Y = 0$$

if $Y = 6$ and $dY/dT = 0$ at $T = 0$.

(b) Find the particular solution of

$$\frac{d^2X}{dT^2} - 4X = 0$$

if $X = 2$ at $T = 0$ and $X = 4$ at $T = 1$.

(Leave e in your answer.)

You may wonder what a solution of a differential equation of the family $d^2y/dx^2 - \lambda^2 y = 0$ looks like graphically. If either p or q is zero, the curve is an exponential of the type you met in Unit 8. Otherwise, the solution

$$y = p\exp\lambda x + q\exp(-\lambda x)$$

behaves like $y = p\exp\lambda x$ when x is large and positive (because then $q\exp(-\lambda x)$ makes only a very small contribution to the value of y) and behaves like $y = q\exp(-\lambda x)$ when x is large and negative (because then $p\exp\lambda x$ makes only a very small contribution to the value of y).

Figure 1 shows the four general shapes of curve you can expect. In Figure 1(a), p and q are both positive; there is a minimum point and y is always positive. In Figure 1(b), p and q are both negative; there is a maximum point and y is always negative. In Figures 1(c) and (d), p and q are of opposite sign; there is a point of inflection in both cases, but when p is positive y increases with increasing x while when p is negative then y decreases with increasing x. So you can see that the shape of the graph of the solution can vary quite widely, depending on p and q.

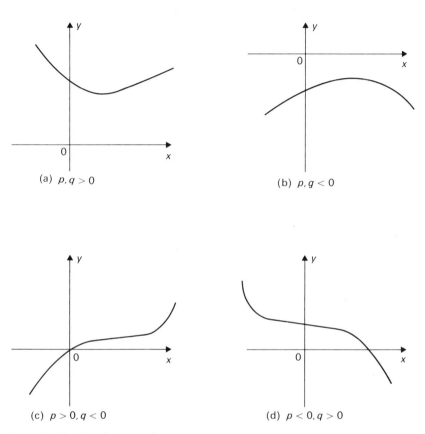

(a) $p, q > 0$

(b) $p, q < 0$

(c) $p > 0, q < 0$

(d) $p < 0, q > 0$

Figure 1   Sketches of $y = p\exp\lambda x + q\exp(-\lambda x)$

3.2   The solution of $d^2y/dx^2 + \lambda^2 y = 0$

Earlier I found that an exponential is not a solution of differential equation (9)

$$\frac{d^2y}{dx^2} + \lambda^2 y = 0 \qquad (\lambda \neq 0)$$

If I rewrite it as

$$\frac{d^2y}{dx^2} = -\lambda^2 y$$

you can see that we need to find a function which, when differentiated twice, is a negative constant times what it was originally.

You may remember from Unit 9 that if you differentiate $\cos\omega t$ or $\sin\omega t$ twice with respect to t you obtain $-\omega^2\cos\omega t$ and $-\omega^2\sin\omega t$, respectively. This suggests that a sine or cosine function could be a solution of differential equation (9).

SAQ 18

Differentiate

(a) $y = p \sin kx$
(b) $y = p \cos kx$

where p and k are constants.

Can either of them be a solution of differential equation (9) under any circumstances?

You will have found that both $y = p \sin kx$ and $y = p \cos kx$ are solutions provided $k^2 = \lambda^2$, that is provided $k = \lambda$. (I shall ignore the case $k = -\lambda$; because $\cos(-kx) = \cos(kx)$ and $\sin(-kx) = -\sin(kx)$ it leads effectively to the same solution.) For the general solution I need a solution with two arbitrary constants.

Can you suggest the form of the general solution?

By analogy, with the general solution in Section 3.1, it is

$$y = p \sin \lambda x + q \cos \lambda x \qquad (12)$$

where p and q are arbitrary constants.

(If you like you can check functions like $y = p \sin \lambda x + bx$. You will find they are not solutions. Only equation (12) has two arbitrary constants and satisfies the differential equation.)

Example
Find the general solution of

$$\frac{d^2 Y}{dX^2} + 9Y = 0$$

In this case, 9 corresponds to λ^2 and so I can write the general solution, equation (12), as

$$Y = P \sin 3X + Q \cos 3X$$

where P and Q are arbitrary constants.

SAQ 19

(a) Write down the general solution of each of the following

(i) $\dfrac{d^2 Y}{dX^2} + 25Y = 0$

(ii) $9\dfrac{d^2 X}{dT^2} + 4X = 0$

(iii) $\dfrac{d^2 y}{dt^2} + \omega^2 y = 0$

(b) Find the particular solution of

$$\frac{d^2 Y}{dX^2} + 16Y = 0$$

given $Y = 0$ at $X = 0$ and $Y = 3$ at $X = \pi/8$.

The graph of the solution of the differential equation $d^2 y/dx^2 + \lambda^2 y = 0$ is always sinusoidal in shape. By *sinusoidal* I mean that it has the shape of a   *sinusoidal* sine or cosine curve. This does not mean it must cut the axes at the same points as would a sine or cosine curve. As the arbitrary constants p and q vary, the peak-to-peak height of the curve varies, as do the points where it cuts the horizontal axis. This does not, however, affect the basic sinusoidal shape, which is shown in Figure 2.

Figure 2 Sketch of $y = p\sin \lambda x + q\cos \lambda x$

SAQ 20

SAQ 20

(a) Find the particular solution of

$$\frac{d^2 Y}{dX^2} + Y = 0$$

given $Y = 0$ and $dY/dX = 3$ when $X = 0$.

(b) Find the particular solution of

$$\frac{d^2 Y}{dT^2} - 4Y = 0$$

given $Y = 0$ and $dY/dT = 4$ when $T = 0$.

3.3 The solution of $ad^2y/dx^2 + bdy/dx + cy = 0$

You have now seen how to write down solutions of the differential equation

$$a\frac{d^2 y}{dx^2} + b\frac{dy}{dx} + cy = 0 \qquad (a \neq 0) \tag{13}$$

when $b = 0$ for the case where a and c have the same sign and for the case where they have opposite signs. You have found that in the first of these cases sinusoids give a solution; in the second exponentials. I shall now turn to the case where $b \neq 0$. Once again, I shall ask you to see if an exponential can give a solution

SAQ 21

SAQ 21

Differentiate $y = p\exp kx$ (p and k constants) twice. By substitution, find the circumstances in which this equation can be a solution of differential equation (13),

$$a\frac{d^2 y}{dx^2} + b\frac{dy}{dx} + cy = 0 \qquad (a \neq 0)$$

You will have found that $y = p\exp kx$ is a solution to the differential equation provided

$$ak^2 + bk + c = 0$$

As you learned in Unit 6, this quadratic equation has solutions

$$k = \frac{-b \pm \sqrt{(b^2 - 4ac)}}{2a}$$

k is only a real number (that is, a normal number) provided $b^2 \geqslant 4ac$. So $p\exp kx$ can be a solution provided k has the value given by the above expression and provided $b^2 \geqslant 4ac$ in differential equation (13).

The solution when $b^2 > 4ac$ is in fact slightly different from the solution when $b^2 = 4ac$, so I shall look at the two cases separately. I shall then try to find a solution for the case where $b^2 < 4ac$.

27

The case where $b^2 > 4ac$

As a specific example, I shall find the general solution of a particular differential equation of this type,

$$\frac{d^2Y}{dX^2} + 5\frac{dY}{dX} + 6Y = 0$$

Here I am working with dimensionless quantities, and $A = 1$, $B = 5$ and $C = 6$.

So is $B^2 > 4AC$?

Yes; $B^2 = 25$, $4AC = 24$.

The expression

$$Y = P \exp KX$$

is a solution if K is either

$$\frac{-5 + \sqrt{(25 - 24)}}{2}$$

or

$$\frac{-5 - \sqrt{(25 - 24)}}{2}$$

That is, $K = -2$ or $K = -3$. So $Y = P \exp(-2X)$ and $Y = P \exp(-3X)$ are solutions.

A general solution needs *two* arbitrary constants. What do you think might be the general solution?

As you might guess by analogy with the general solutions in Sections 3.1 and 3.2, the general solution is

$$Y = P \exp(-2X) + Q \exp(-3X)$$

where P and Q are arbitrary constants. (You may like to check for yourself that this expression is a solution.)

SAQ 22

SAQ 22

Find the general solutions of the following differential equations.

(a) $\dfrac{d^2Y}{dX^2} + 3\dfrac{dY}{dX} + 2Y = 0$

(b) $6\dfrac{d^2Y}{dT^2} + 17\dfrac{dY}{dT} + 12Y = 0$

I can now give a formula for the general solution of

$$a\frac{d^2y}{dx^2} + b\frac{dy}{dx} + cy = 0 \qquad (a \neq 0)$$

for the case where $b^2 > 4ac$. This formula is given in your *Handbook*, so there is no need for you to memorize it, although you need to know where to look it up and how to use it.

The formula is

$$y = p \exp\left\{\frac{-b + \sqrt{(b^2 - 4ac)}}{2a}x\right\} + q \exp\left\{\frac{-b - \sqrt{(b^2 - 4ac)}}{2a}x\right\}$$

where p and q are arbitrary constants. It looks rather formidable, and you will probably find it easier to come to terms with it if you remember that the coefficients of x in the two exponential terms are the roots of the quadratic equation

$$ak^2 + bk + c = 0$$

as you discovered in SAQ 21.

There is a common factor $\exp(-bx/2a)$ in this formula and so you might sometimes meet it written as

$$y = \exp\left(\frac{-bx}{2a}\right)\left[p\exp\left\{\frac{\sqrt{(b^2 - 4ac)}}{2a}x\right\} + q\exp\left\{\frac{-\sqrt{b^2 - 4ac}}{2a}x\right\}\right]$$

I shall not use this form in this course.

There is a greater variety of shapes which can arise when these solutions are plotted graphically than in the case of the solutions of $d^2y/dx^2 - \lambda^2 y = 0$ or $d^2y/dx^2 + \lambda^2 y = 0$. This is because it is possible to have either two negative exponentials or two positive exponential or one of each. Because the arbitrary constants may also each be positive, negative or zero and they may be equal or unequal, there is a total of over fifty differing cases to consider! One point all the solutions have in common, however, is that, while they may have a maximum or minimum at intermediate values of x, at both large positive and large negative values of x they will be recognizably exponential in shape. Just as an example, Figure 3 shows the graph of $Y = P\exp(-2X) + Q\exp(-3X)$ (which was the solution to the differential equation in my example just before SAQ 22) for the case when $P = 1$ and $Q = -2$.

The case where $b^2 = 4ac$

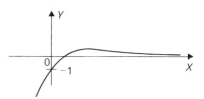

Figure 3   Graph of
$Y = \exp(-2X) - 2\exp(-3X)$

Study comment

This case is included for completeness in my discussion of the solutions of differential equation (13)

$$a\frac{d^2y}{dx^2} + b\frac{dy}{dx} + cy = 0 \qquad (a \neq 0)$$

It is not often encountered in modelling, and it will be sufficient if you simply read through this subsection and note the result (which is quoted in your *Handbook* should you need to refer to it). No exercises are set on this case, and there will be no continuous assessment or examination questions on it.

When you tried a solution to differential equation (13) of the form $y = p\exp kx$ in SAQ 21 you found that

$$k = \frac{-b \pm \sqrt{(b^2 - 4ac)}}{2a}$$

Clearly, should b^2 equal $4ac$, then $k = -b/2a$. Hence you might expect that

$$y = p\exp\left(\frac{-bx}{2a}\right)$$

is a solution of the differential equation and indeed it is. The problem is that it cannot be the general solution because it does not contain two arbitrary constants, yet there is no conveniently 'paired' solution to go with it to make up the general solution. It turns out that the general solution is

$$y = \exp\left(\frac{-bx}{2a}\right)(px + q)$$

where p and q are arbitrary constants.

If you are interested, you might like to check by differentiation that this is indeed the solution if $b^2 = 4ac$ in differential equation (13).

The case where $b^2 < 4ac$

You saw that an exponential was only a solution if the parameters in the differential equation were such that $b^2 \geq 4ac$. So what happens if $b^2 < 4ac$?

Because a sinusoid was the solution the last time an exponential was not (Section 3.2), perhaps a sinusoid can help here.

Unfortunately, if you were to try

$$y = p \sin kx$$

or

$$y = p \cos kx$$

you would find that neither of them is a solution of differential equation (13) under any circumstances.

However, an expression of the form

$$y = p \exp kx \sin mx \qquad (14)$$

is a solution provided k and m are chosen appropriately. In the next SAQ you will investigate how k and m are related to the constant parameters a, b and c in the differential equation.

Study comment
The algebra becomes rather messy in the next SAQ, although the SAQ itself is not difficult. You may prefer to use the SAQ and its answer as a piece of text rather than to work through it yourself. Alternatively, you could pick up the SAQ either at part (c) or at part (d) and concentrate on finding k and m rather than on practising differentiation.

SAQ 23   SAQ 23

(a) Show that if y is given by equation (14), then

$$\frac{dy}{dx} = p \exp kx \, (k \sin mx + m \cos mx)$$

(b) Show that if dy/dx is given by the expression in (a) above then

$$\frac{d^2y}{dx^2} = p \exp kx \{(k^2 - m^2) \sin mx + 2km \cos mx\}$$

(c) Substitute into differential equation (13) the expressions for y, dy/dx and d^2y/dx^2 given by equation (14) and (a) and (b) above. Collect together all the terms which have $\cos mx$ in them and all the terms which have $\sin mx$ in them.

(d) If equation (14) is indeed a solution of this differential equation then your expression must be zero for every possible value of x. Because $\exp kx$ is never quite zero for any finite value of x and because $\sin mx$ and $\cos mx$ are only zero for specific values of x, the only way the expression you have found can be zero for every possible value of x is for both the coefficient of $\sin mx$ and the coefficient of $\cos mx$ to be zero. Show that this leads to

$$k = \frac{-b}{2a} \quad \text{and} \quad m = \frac{\pm \sqrt{(4ac - b^2)}}{2a}$$

(*Hint:* use the coefficient of $\cos mx$ first.)

You have now found that equation (14)

$$y = p \exp kx \sin mx$$

is a solution provided that

$$k = \frac{-b}{2a} \quad \text{and} \quad m = \frac{\pm\sqrt{(4ac - b^2)}}{2a}$$

Another solution is

$$y = q \exp kx \cos mx$$

with the same values for k and m. As you might guess, the *general* solution is

$$y = \exp kx(p \sin mx + q \cos mx)$$

with the same values for k and m. I can ignore the negative value for m, just as I did in Section 3.2, because $\sin(-mx) = -\sin mx$ and $\cos(-mx) = \cos mx$. Hence, the general solution is

$$y = \exp\left(\frac{-bx}{2a}\right)\left[p \sin\left\{\frac{\sqrt{(4ac - b^2)}}{2a}x\right\} + q \cos\left\{\frac{\sqrt{(4ac - b^2)}}{2a}x\right\}\right]$$

where p and q are arbitrary constants.

Once again, there is no need to memorize this solution; it is given in your *Handbook*. Figure 4 shows the form of the solution when $-b/2a$ is positive and also when it is negative. You can see that the effect of multiplying the exponential and sinusoid together is to produce either decaying or growing oscillations. (You might well be able to think straight away of some situations such a differential equation will model, just from the shape of the solution. You will meet some models using this equation in Unit 15.)

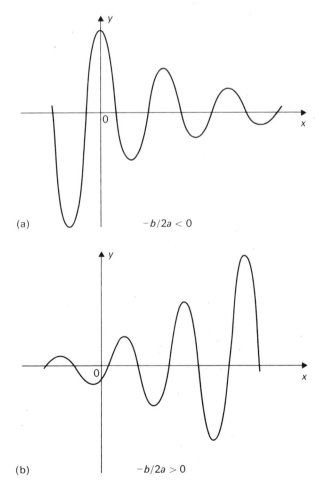

(a) $-b/2a < 0$

(b) $-b/2a > 0$

Figure 4 Sketches of the solutions of a $d^2y/dx^2 + b\,dy/dx + cy = 0$ when $b^2 < 4ac$.

Example

Find the general solution of

$$\frac{d^2Y}{dX^2} + 3\frac{dY}{dX} + 5Y = 0$$

Here I am working with dimensionless quantities and $A = 1$, $B = 3$ and $C = 5$.

So is $B^2 < 4AC$?

$B^2 = 9$ and $4AC = 20$, so B^2 is indeed less than $4AC$.

The general solution here is of the form

$$Y = \exp\left(\frac{-BX}{2A}\right)\left[P\sin\left\{\frac{\sqrt{(4AC - B^2)}}{2A}X\right\} + Q\cos\left\{\frac{\sqrt{(4AC - B^2)}}{2A}X\right\}\right]$$

I need to evaluate $-B/2A$ and $\sqrt{(4AC - B^2)}/2A$.

$$\frac{-B}{2A} = \frac{-3}{2}$$

$$= -1.5$$

$$\frac{\sqrt{(4AC - B^2)}}{2A} = \frac{\sqrt{(20 - 9)}}{2}$$

$$= \frac{\sqrt{11}}{2}$$

$$= 1.66 \quad \text{(to three significant figures)}$$

Hence in this case, the general solution is

$$Y = \exp(-1.5X)(P\sin 1.66X + Q\cos 1.66X)$$

where P and Q are arbitrary constants.

Notice how the solution looks very much less messy as soon as numbers replace A, B and C in the exponential, sine and cosine terms.

SAQ 24

SAQ 24

Find the general solutions of the following differential equations

(a) $\dfrac{d^2Y}{dX^2} + 2\dfrac{dY}{dX} + 2Y = 0$

(b) $\dfrac{d^2Y}{dX^2} + 4\dfrac{dY}{dX} + 5Y = 0$

All three cases

I have now dealt with all three of the possible cases which can arise. Let me summarize the results.

1   If $b^2 > 4ac$, the solution is the sum or difference of two exponentials, and has the form

$$y = p\exp\left(\frac{-b + \sqrt{(b^2 - 4ac)}}{2a}x\right) + q\exp\left(\frac{-b - \sqrt{(b^2 - 4ac)}}{2a}x\right)$$

where p and q are arbitrary constants.

2   If $b^2 = 4ac$, the solution is an exponential multiplying a linear term, and has the form

$$y = \exp\left(\frac{-bx}{2a}\right)(px + q)$$

where p and q are arbitrary constants.

3   If $b^2 < 4ac$, the solution is a decaying or growing sinusoid, and has the form

$$y = \exp\left(\frac{-bx}{2a}\right)\left[p\sin\left\{\frac{\sqrt{(4ac - b^2)}}{2a}x\right\} + q\cos\left\{\frac{\sqrt{(4ac - b^2)}}{2a}x\right\}\right]$$

where p and q are arbitrary constants.

So to find the general solution of a second-order differential equation of the type

$$a\frac{d^2y}{dx^2} + b\frac{dy}{dx} + cy = 0 \qquad (a \neq 0)$$

(1) check the relative values of a, b and c,

(2) select the form of the solution using your *Handbook*,

(3) evaluate $-b/2a$ and $\dfrac{\sqrt{(b^2 - 4ac)}}{2a}$ or $\dfrac{\sqrt{(4ac - b^2)}}{2a}$ for the given differential equation,

(4) write down the general solution.

SAQ 25 SAQ 25

Find the general solution of each of the following differential equations.

(a) $\dfrac{d^2Y}{dT^2} + 6\dfrac{dY}{dT} + 8Y = 0$

(b) $\dfrac{d^2X}{dT^2} + 6\dfrac{dX}{dT} + 10X = 0$

Particular solutions can be found using initial or boundary conditions. There is an example of this on side 2 of Disc 8.

Study comment
You should listen to both sides of Disc 8 now.

Does this method of comparing b^2 with $4ac$ and then selecting the form of the solution work even if $b = 0$, that is for the equations introduced in Section 3.1 and 3.2, $d^2y/dx^2 + \lambda^2 y = 0$ and $d^2y/dx^2 - \lambda^2 y = 0$?

Yes, it does. Looking at $d^2y/dx^2 - \lambda^2 y = 0$ first, here $b = 0$, $a = 1$ and $c = -\lambda^2$, so $b^2 = 0$ and $4ac = -4\lambda^2$. Hence, $b^2 > 4ac$ and the solution is of the form (case 1 above):

$$y = p\exp\left\{\frac{-b + \sqrt{(b^2 - 4ac)}}{2a}x\right\} + q\exp\left\{\frac{-b - \sqrt{(b^2 - 4ac)}}{2a}x\right\}$$

Putting in values for a, b and c gives

$$y = p\exp\lambda x + q\exp(-\lambda x)$$

which is the result obtained in Section 3.1.

Similarly, for $d^2y/dx^2 + \lambda^2 y = 0$: $b = 0$, $a = 1$ and $c = \lambda^2$, so $b^2 < 4ac$, and case 3 above shows

$$y = \exp\left(\frac{-bx}{2a}\right)\left[p\sin\left\{\frac{\sqrt{(4ac - b^2)}}{2ac}x\right\} + q\cos\left\{\frac{\sqrt{(4ac - b^2)}}{2a}x\right\}\right]$$

Putting in values for a, b and c gives

$$y = p\sin\lambda x + q\cos\lambda x$$

which is the result obtained in Section 3.2.

So, although I chose to introduce these two differential equations separately, they could be treated in just the same way as any member of the family of differential equations

$$a\frac{d^2y}{dx^2} + b\frac{dy}{dx} + cy = 0 \qquad (a \neq 0)$$

but, because their solutions are simpler, it is unnecessarily cumbersome to use the general form of solution for them.

3.4 Summary

This section has introduced the general solutions of all the members of the family of differential equations

$$a\frac{d^2y}{dx^2} + b\frac{dy}{dx} + cy = 0 \qquad (a \neq 0) \tag{13}$$

These solutions are sums or differences of exponentials, or sinusoids, or products of an exponential and sinusoids or linear functions. They all contain two arbitrary constants.

When $b = 0$, and $c/a < 0$, then the differential equation may be written

$$\frac{d^2y}{dx^2} - \lambda^2 y = 0 \tag{10}$$

and has the general solution

$$y = p\exp\lambda x + q\exp(-\lambda x) \tag{11}$$

where p and q are arbitrary constants.

When $b = 0$ and $c/a > 0$ then the differential equation may be written

$$\frac{d^2y}{dx^2} + \lambda^2 y = 0 \tag{9}$$

and has the general solution

$$y = p\sin\lambda x + q\cos\lambda x \tag{12}$$

where p and q are arbitrary constants.

When $b \neq 0$, the solution depends on the relative values of a, b and c. The three possible cases are summarized on pages 32 and 33 of this unit and in your *Handbook*.

4 A MODEL OF HEAT FLOW DOWN A FIN

In this section I want to show you how the second-order differential equation

$$\frac{d^2y}{dx^2} - \lambda^2 y = 0$$

can be used to model a problem in heat flow. This problem has to do with the cooling of a motorcycle engine by the use of cooling fins. If you look at the photograph in Figure 5 you can see the fins which are used in air-cooling the engine. Fins are used because the surface area of the engine alone is not large enough to enable all the waste heat produced by the engine to be removed.

Figure 5   A motorcycle engine, showing the fins

These fins increase the surface area over which cooling takes place; but they also add to the cost of the engine, and so the manufacturers will want to ensure that they are adequate without being any larger than necessary.

If, therefore, an enthusiastic owner of the motorcycle wants to make some adjustments to his engine to increase its performance, he will need to be sure that the existing fins can cope with the new, increased, waste heat output. How can he do this?

To answer this question I shall set up a model.

4.1   Setting up a model to describe the steady flow of heat along a fin

Study comment
You do not need to reproduce any of the facts about heat in this section or the arguments leading to differential equation (22). You should, however, follow the arguments to see where the differential equation comes from—and you should be able to solve the differential equation.

First, I must make quite clear what I am modelling. It is the heat loss through the cooling fins of a hot engine. I shall suppose the fins to be rectangular, as in Figure 6. Each fin will lose heat through its sides, top, bottom and outer edge; that is, from the faces corresponding to ABCD, EFGH, ADEH, BCFG and CDEF in Figure 6. The heat loss through the top and bottom faces (ADEH and BCFG) will be affected by the proximity of the neighbouring fins, but I shall ignore this in my model; a supposition which is not too bad when the motor cycle is moving and air is being forced past the fins. I shall also ignore all heat loss except that through the fins. This is a 'safe' supposition, because it will lead to an *under*estimate of the heat which can be safely dissipated.

heat source

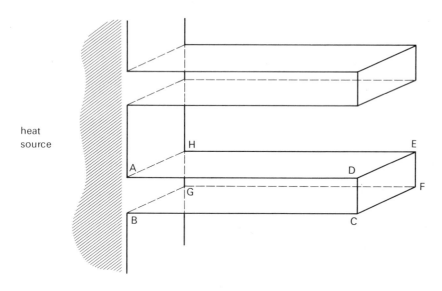

Figure 6   Idealization of the cooling fins on an engine (thickness exaggerated for clarity)

Second, I need to be clear about the purpose of the model. It is to find whether the fins can provide effective cooling when the engine's performance, and therefore the waste heat output, is increased.

Finally, I need to be clear about which factors are known for the engine. These are the rate at which waste heat will be produced by the modified engine, the maximum safe operating temperature of the engine and the number and size of the fins. Also known are physical constants, such as thermal conductivities and heat transfer coefficients.

In my model I shall look at the situation where there is no time variation in the heat flow along the fins—the so-called 'steady state' when the engine is hot, the fins are warm and all the waste heat is being dissipated to the surrounding air.

To simplify things, I shall suppose that all the fins are identical, and that, therefore, I can examine the heat flow through just one fin of length l, width w and thickness b. I shall define a variable x to be the distance along the fin measured from the engine. Thus, the fin extends from $x = 0$, the inner end to $x = l$, the outer end. Figure 7 summarizes these details.

As heat flows out from the hot engine and along the fin, it escapes through the top, the bottom and the sides.

Would you expect the heat flow to be constant along the length of the fin?

The sideways losses must reduce the flow along the length of the fin; that is, from left to right in Figure 7.

Thus, I should use a model which allows for a variation of q with x, where q is the rate of heat flow. Similarly, I should use a model which allows the temperature of the fin θ to vary with the distance along the fin x.

Figure 7 shows that, if I consider a thin slab between $x = x_1$ and $x = x_2$, heat flows in at a rate q_1, heat flows out at a rate q_2 along the fin and heat flows out of the sides of the slab at a rate q_L. In the steady state, once the engine has warmed to its normal operating temperature and the fin has also warmed up, q_1, q_2 and q_L will not vary with time.

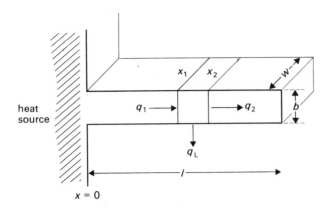

Figure 7   *The fin is of length l, width w and thickness b; a thin slab between x_1 and x_2 is considered.*

How do you expect q_1, q_2 and q_L to be related?

The rate of heat flow into the slab equals the rate of heat flow out, thus

$$q_1 = q_2 + q_L \tag{15}$$

This key equation represents the basic idea that the rate at which heat flows into the slice equals the rate at which heat leaves it; no heat is stored and none vanishes, thus the temperature is constant. The equation therefore applies once the steady state has been reached.

In Unit 13 an equation was introduced relating the rate of heat loss from a hot object to the surface area of the hot object and the temperature difference between the object and the surrounding air. (See Unit 13, Section 2.2.) This equation was

$$q = ja\theta \tag{16}$$

where j is the heat transfer coefficient, a the area of the surface and θ the temperature difference.

I can use this equation to find an expression for q_L, the rate of heat loss through the sides of the slab, in terms of dimensions of the slab and its temperature. However, the expression will only be an approximation.

Can you see why?

You already know that the temperature difference θ is different for different values of x. Thus, I should have to use an *intermediate* value for θ between x_1 and x_2 when using equation (16). Of course, the closer x_1 is to x_2, the better the approximation.

What should I use here for a in equation (16)?

a is the surface area. Hence, it is equal to $2(x_2 - x_1)w$ for the top and bottom faces, plus $2(x_2 - x_1)b$ for the side faces. Thus

$$a = 2(w + b)(x_2 - x_1)$$

and equation (16) becomes

$$q_L = 2j(w + b)(x_2 - x_1)\theta$$

37

Instead of q_L I can write, from equation (15), $q_1 - q_2$, thus obtaining

$$q_1 - q_2 = 2j(w + b)(x_2 - x_1)\theta \tag{17}$$

I shall now let x_2 and x_1 be very close. I will show that this is the case in my equations by putting

$$x_1 = x$$

and $\qquad x_2 = x + \Delta x$

Because x_1 and x_2 are nearly equal, q_1 and q_2 will also be nearly equal. I shall refer to q_1 as q, therefore, and to q_2 as $q + \Delta q$. Putting all this in equation (17) gives

$$q - (q + \Delta q) = 2j(w + b)(x + \Delta x - x)\theta$$

which is

$$-\Delta q = 2j(w + b)\theta \, \Delta x$$

or

$$\frac{\Delta q}{\Delta x} = -2j(w + b)\theta$$

In the limit, as Δx tends to zero, $\Delta q/\Delta x$ tends to dq/dx and I get

$$\frac{dq}{dx} = -2j(b + w)\theta \tag{18}$$

When obtaining a new equation it is often useful to examine the signs of each term to check that they make sense. In equation (18), j and $(w + b)$ are positive constants, θ is also positive because the fin is hotter than its surroundings. Equation (18) therefore tells me that the derivative of q with respect to x is negative.

Is the reasonable?

Yes, because the rate of flow of heat along the fin in the direction of increasing x is being *reduced* by loss of heat from the sides.

Can differential equation (18) be solved?

There is a problem. It is that, although j, w and b are independent of x, θ depends on x. That is, I have too many dependent variables; I should have either q or θ, but not both.

What I must do is look for some other relationship between some or all of q, θ and x. If I can find one, I can substitute into equation (18) to remove the excess variable.

One of the equations introduced in Unit 13 can help here. It is the equation for the rate of flow of heat through a wall (see Unit 13, Section 3.1):

$$q = \frac{\kappa a(\theta_i - \theta_o)}{d} \tag{19}$$

where κ is the thermal conductivity of the wall, a the area of the wall, d its thickness and θ_i and θ_o the temperatures of the inner and outer faces, respectively.

Now in Unit 13, q did not vary with distance through the wall. In the present problem it does vary because q decreases along the length of the fin. So I cannot use equation (19) as it stands for the whole fin. But I can use the equation for a thin slab of fin; the slab which is shown in Figure 7 and which I used to derive equation (18).

I shall let the temperature of the slab be θ_1 at $x = x_1$ and θ_2 at $x = x_2$. Then, in this case equation (19) becomes

$$q = \frac{\kappa wb(\theta_1 - \theta_2)}{x_2 - x_1} \tag{20}$$

because the area here is the area of cross-section of the fin wb. Equation (20) is an approximation.

Why?

The equation could only be exact if q was the same at x_1 as at x_2. But you have just seen that q varies with x. Hence, the value of q in equation (20) must be an approximation—an intermediate value of q between that at x_1 and that at x_2.

I shall again let x_1 be very close to x_2 and write $x_1 = x$ and $x_2 = x + \Delta x$. I shall also write θ_1 as θ, θ_2 as $\theta + \Delta\theta$. Then equation (20) becomes

$$q = \frac{\kappa wb\{\theta - (\theta + \Delta\theta)\}}{(x + \Delta x) - x}$$

This is a better approximation than equation (20) because Δx is small and so q will scarcely vary between the two values of x. The equation simplifies to

$$q = \frac{-\kappa wb\,\Delta\theta}{\Delta x}$$

and in the limit as $\Delta x \longrightarrow 0$, $\dfrac{\Delta\theta}{\Delta x} \longrightarrow \dfrac{d\theta}{dx}$ and the equation becomes

$$q = -\kappa wb\frac{d\theta}{dx} \tag{21}$$

Once again, let us check the sign in this new differential equation. κ, w and b are all positive constants, so the equation tells us that q can only be positive if $d\theta/dx$ is negative.

Is $d\theta/dx$ negative?

Yes, θ decreases as x increases towards the tip of the fin.

Equation (21) gives me the second relationship among q, x and θ that I needed after deriving equation (18)

$$\frac{dq}{dx} = -2j(w + b)\theta \tag{18}$$

Can you see how I can arrive at a single differential equation relating q to x from equations (18) and (21)?

I can eliminate θ from these two equations as follows. First, I differentiate equation (18) with respect to x. j, w and b are all independent of x, so the result is

$$\frac{d^2q}{dx^2} = -2j(w + b)\frac{d\theta}{dx}$$

Next I substitute for $d\theta/dx$ using equation (21), to obtain

$$\frac{d^2q}{dx^2} = \frac{-2j(w + b)}{-\kappa wb}q$$

39

which can be tidied up to give

$$\frac{d^2q}{dx^2} = \frac{2j(w+b)}{\kappa wb}q \tag{22}$$

This, at last, is a differential equation I can solve. Because j, w, b and κ are all positive, it is like one of the second-order differential equations introduced in Section 3.3, differential equation (10):

$$\frac{d^2y}{dx^2} - \lambda^2 y = 0$$

with y replaced by q and λ^2 by $\dfrac{2j(w+b)}{\kappa wb}$. It is, in fact, the differential equation I am looking for; it relates rate of heat flow to distance along the bar. If I have some initial or boundary conditions and know values for j, w, b and κ then I have enough information to answer my question: can the fins cool the modified engine?

4.2 Finding a particular solution of the differential equation in the model

SAQ 26

SAQ 26

Find the general solution of differential equation (22).

The solution contains two arbitrary constants. It is an equation I can use to model the heat flow down any fin. With sufficient information about a particular engine, I can find initial or boundary conditions which, when used with the general solution, will enable me to specify whether the fins can cool that engine after it has been modified.

In my problem, the engine has the specification shown in Table 1. It is expected that the modification will increase the waste heat output by about 25 per cent.

Table 1

Parameter	Value for engine
Thermal conductivity of material of fins (aluminium) (κ)	$201\ \text{W m}^{-1}\text{K}^{-1}$
Heat transfer coefficient from fins (cycle moving) (j)	$25\ \text{W m}^{-2}\text{K}^{-1}$
Waste heat output rate from *unmodified* engine	
Maximum operating temperature of engine (*excess* over environment temperature)	$200\ \text{K}$
Length of each fin (l)	$5\,\text{cm} = 0.05\ \text{m}$
Width of each fin (w)	$20\,\text{cm} = 0.2\ \text{m}$
Thickness of each fin (b)	$1\,\text{mm} = 0.001\ \text{m}$
Number of fins	20

First, notice that there is enough information in this table to specify the constant

$$\sqrt{\frac{2j(b+w)}{\kappa bw}}\,\Big\}$$

which occurs in the general solution you found in SAQ 26,

$$q = p \exp\left[\sqrt{\left\{\frac{2j(b+w)}{\kappa bw}\right\}}x\right] + m \exp\left[-\sqrt{\left\{\frac{2j(b+w)}{\kappa bw}\right\}}x\right] \qquad (23)$$

where p and m are arbitrary constants. (I am using m instead of q here because q already appears in the equation.)

The constant has the value

$$\sqrt{\left|\left(\frac{2 \times 25 \times 0.201}{201 \times 0.2 \times 0.001}\right)\right|}$$

$$= \sqrt{250}$$

$$= 15.8 \quad \text{(to three significant figures)}$$

I can therefore write the general solution in the dimensionless form

$$Q = P \exp 15.8X + M \exp(-15\cdot8X) \qquad (24)$$

My next step is to find initial or boundary conditions from the information available in Table 1.

Can you see any relevant data?

I need information about Q or dQ/dX at particular values of X.

I have one such value; it is the expected value of the waste heat output rate after the engine is modified. Because an increase of 25 per cent is expected, this rate is 1250 W.

For what value of X can this value of Q be used?

It is applicable to $X = 0$.

However, my condition is *not* $Q = 1250$ at $X = 0$ because I am modelling the heat flow down just one of 20 fins.

What is the correct value of Q at $X = 0$ for each fin?

It is $\frac{1250}{20} = 62.5$, because each of the fins takes $\frac{1}{20}$ of the output. (Remember, my model supposes that *all* heat loss is through the fins.)

This is one condition. For the other condition I need either a value of Q at another value of X or a value of dQ/dX at some value of X.

Have I got such information?

No

It appears that I shall have to make a supposition to obtain the other condition. I shall do this by supposing that the fin is sufficiently long that at $X = L$ (the tip of the fin) no heat is lost through the tip because it has all been lost already through the sides. Because the fin's tip has a small area this may not be too inaccurate a supposition. It is also a 'safe' supposition because if heat *is* lost through the tip the engine will be more efficiently cooled than predicted by the model. It certainly enables me to write down a second condition, which is $Q = 0$ at $X = L = 0.05$.

Are the two conditions just derived initial or boundary conditions?

They are boundary conditions, two values of Q at two values of X.

I can use these two boundary conditions to arrive at the particular solution for my engine. The general solution, equation (24), was

$$Q = P \exp 15.8X + M \exp(-15 \cdot 8X)$$

Putting $Q = 62.5$ at $X = 0$ gives

$$62.5 = P \exp 0 + M \exp 0$$
$$= P + M \qquad (25)$$

Putting $Q = 0$ at $X = 0.05$ gives

$$0 = P \exp(15.8 \times 0.05) + M \exp(-15.8 \times 0.05)$$
$$= P \exp 0.79 + M \exp(-0.79)$$

I can multiply both sides by $\exp 0.79$ to obtain

$$0 = P \exp 1.58 + M \qquad (26)$$

Substituting for M from equation (25) gives

$$0 = P \exp 1.58 + 62.5 - P$$

so

$$P(1 - \exp 1.58) = 62.5$$

and

$$P = \frac{62.5}{1 - \exp 1.58}$$
$$= \frac{62.5}{1 - 4.855}$$
$$= -16.2 \quad \text{(to three significant figures)}$$

Putting this in equation (26) gives

$$M = \frac{-62.5 \exp 1.58}{1 - \exp 1.58}$$
$$= \frac{-62.5 \times 4.855}{1 - 4.855}$$
$$= 78.7 \quad \text{(to three significant figures)}$$

Hence the particular solution is

$$Q = -16.2 \exp 15.8X + 78.7 \exp(-15.8X) \qquad (27)$$

How can I use this equation to check that the engine will not overheat?

There is one further piece of information in Table 1 which I have not yet used. This is the maximum safe operating temperature excess of the engine. If I could use my model to find a value of the temperature excess θ at $X = 0$, I could then check if it was below this maximum value.

If I know an expression for Q, can I find one for θ?

Yes; I need to use equation (18)

$$\frac{\mathrm{d}q}{\mathrm{d}x} = -2j(w + b)\theta$$

I can differentiate equation (27), and then divide it by $-2j(w + b)$ to obtain an expression for θ. Putting $X = 0$ will then give the predicted value of θ for the modified engine.

SAQ 27

Is the engine safe when modified?

SAQ 27

So my model has answered my question. It looks as if the modification will not take the operating temperature up to the maximum safe value.

4.3 Another use of the model

Motor cycle engines are not the only devices cooled by fins. A power transistor can also be cooled in this way. Consider a particular case: Suppose the fin for a power transistor is 10 cm long, 10 cm wide and 1 mm thick. It is made of copper, whose thermal conductivity κ is $385 \, \text{W m}^{-1} \, \text{K}^{-1}$. The heat transfer coefficient j between the fin surface and the air is $5.0 \, \text{W m}^{-2} \, \text{K}^{-1}$. If the transistor keeps one edge of the fin at 100 K above the air temperature, what will the excess temperature of the remote edge be and how will the temperature vary along the fin?

The model is rather rough in this case, because a transistor does not heat a straight edge of the fin, but the model gives some idea of the likely temperatures involved.

The first thing to notice here is that the boundary conditions are different. At $x = 0$ it is θ and *not* q that is known. Therefore, I shall have to return to the general solution of the differential equation and use the new boundary conditions.

What second boundary condition can be specified?

Once again, a reasonable supposition is $q = 0$ at $x = l$ where l is the length of the fin.

The general solution is given by equation (23)

$$q = p \exp\left[\sqrt{\left\{ \frac{2j(b + w)}{\kappa b w} \right\}} x \right] + m \exp\left[-\sqrt{\left\{ \frac{2j(b + w)}{\kappa b w} \right\}} x \right]$$

where p and m are arbitrary constants.

This time I shall not put numbers in at this stage. I shall refer to the rather cumbersome expression $\sqrt{\left\{ \frac{2j(b + w)}{\kappa b w} \right\}}$ as λ. I shall denote by θ_0 the temperature of the fin at $x = 0$.

Then the general solution is

$$q = p \exp \lambda x + m \exp(-\lambda x) \tag{28}$$

One condition is $q = 0$ at $x = l$. This gives

$$0 = p \exp \lambda l + m \exp(-\lambda l)$$

which can be multiplied through by $\exp \lambda l$ to give

$$0 = p \exp 2\lambda l + m$$

or

$$m = -p \exp 2\lambda l \tag{29}$$

The other condition is $\theta = \theta_0$ at $x = 0$. Because θ is proportional to dq/dx, I shall have to differentiate equation (28) in order to use this condition.

$$\frac{dq}{dx} = \lambda p \exp \lambda x - \lambda m \exp(-\lambda x)$$

from equation (18),

$$\theta = \frac{-1}{2j(w + b)} \frac{dq}{dx}$$

$$= \frac{-1}{2j(w + b)} \{\lambda p \exp \lambda x - \lambda m \exp(-\lambda x)\} \tag{30}$$

Putting $\theta = \theta_0$ at $x = 0$ gives

$$\theta_0 = \frac{-\lambda}{2j(w + b)} (p - m) \tag{31}$$

Equations (29) and (31) are simultaneous equations for p and m. I can substitute for m from equation (29) into equation (31) to give

$$\theta_0 = \frac{-\lambda}{2j(w+b)}(p + p\exp 2\lambda l)$$

so

$$\frac{-2j(w+b)\theta_0}{\lambda} = p(1 + \exp 2\lambda l)$$

and

$$p = \frac{-2j(w+b)\theta_0}{\lambda(1 + \exp 2\lambda l)}$$

Hence, using equation (29)

$$m = \frac{2j(w+b)\theta_0 \exp 2\lambda l}{\lambda(1 + \exp 2\lambda l)}$$

I want to know the value of θ when $x = l$ and also how θ varies with l along the fin. Hence, I need to substitute these values of p and m into equation (30), the equation for θ. Doing this gives

$$\theta = \frac{-1}{2j(w+b)}\left\{\frac{-2j\lambda(w+b)\theta_0 \exp \lambda x}{\lambda(1 + \exp 2\lambda l)} - \frac{2j\lambda(w+b)\theta_0 \exp 2\lambda l \exp(-\lambda x)}{\lambda(1 + \exp 2\lambda l)}\right\}$$

Tidying this expression up gives

$$\theta = \frac{\theta_0 \exp \lambda x}{1 + \exp 2\lambda l} + \frac{\theta_0 \exp 2\lambda l \exp(-\lambda x)}{1 + \exp 2\lambda l}$$

Now, $\exp 2\lambda l \exp(-\lambda x) = \exp(2\lambda l - \lambda x)$ and there is also a common factor $\theta_0/(1 + \exp 2\lambda l)$. So this equation can be tidied up to give

$$\theta = \frac{\theta_0\{\exp \lambda x + \exp(2\lambda l - \lambda x)\}}{1 + \exp 2\lambda l} \tag{32}$$

SAQ 28 SAQ 28

Use equation (32) and the data given at the beginning of Section 4.3 to find the excess temperature at the remote edge of the fin.

Equation (32) also answers the second question posed at the beginning of this section: how does the temperature vary along the fin? The equation describes the temperature variation.

Using the values for λ and $\exp 2\lambda l$ you will have evaluated in SAQ 32 gives the expression

$$\theta = \frac{100\{\exp 5.12X + \exp(1.024 - 5.12X)\}}{3.784}$$

$$= 26.4\{\exp 5.12X + \exp(1.024 - 5.12X)\}$$

You can see that the variation of temperature is represented by the sum of two exponentials, one increasing with X and the other decreasing as X increases. Figure 8 shows the result in graphical form. The decreasing exponential dominates, so the temperature decreases along the bar.

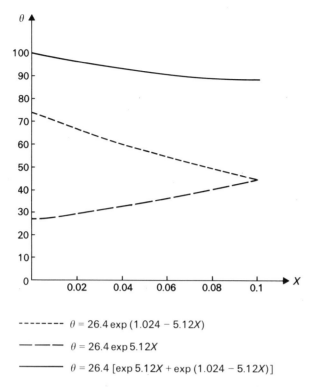

$\quad\text{-------}\quad \theta = 26.4 \exp(1.024 - 5.12X)$

$\quad\text{— —— —}\quad \theta = 26.4 \exp 5.12X$

$\quad\text{————}\quad \theta = 26.4\,[\exp 5.12X + \exp(1.024 - 5.12X)]$

Figure 8   The graph of $\theta = 26.4\{\exp 5.12X + \exp(1.024 - 5.12X)\}$

You have now seen examples of how a second-order differential equation of the form $\mathrm{d}^2y/\mathrm{d}x^2 - \lambda^2 y = 0$ can be used in modelling. There is not space here to cover other examples of models where an equation of this type occurs. Such models usually have to do with the diffusion of a substance and indeed this differential equation is sometimes called the diffusion equation. You may meet these models in other Open University Science or Technology courses.

SUMMARY OF THE UNIT

This unit discusses *second-order differential equations* of two types: Section 1

$$\mathrm{d}^2y/\mathrm{d}t^2 = f(t)$$

and

$$a\,\mathrm{d}^2y/\mathrm{d}x^2 + b\,\mathrm{d}y/\mathrm{d}x + cy = 0.$$

The solution to $\mathrm{d}^2y/\mathrm{d}t^2 = f(t)$ can be found by integrating $f(t)$ twice with   Section 2
respect to t. The general solution will have *two* arbitrary constants. Values
of these constants can be found in particular instances by using *initial
conditions* or *boundary conditions*. Initial conditions give a value of y and a
value of $\mathrm{d}y/\mathrm{d}t$ for some particular value of t. Boundary conditions give
values of y or of $\mathrm{d}y/\mathrm{d}t$ for different values of t. To find values for the arbitrary
constants it is often necessary to solve simultaneous equations that have
been obtained from the initial or boundary conditions.

The differential equations   Section 3

$$\frac{\mathrm{d}^2y}{\mathrm{d}x^2} + \lambda^2 y = 0 \tag{9}$$

and

$$\frac{\mathrm{d}^2y}{\mathrm{d}x^2} - \lambda^2 y = 0 \tag{10}$$

are particular examples of a general type of differential equation

$$a\frac{\mathrm{d}^2y}{\mathrm{d}x^2} + b\frac{\mathrm{d}y}{\mathrm{d}x} + cy = 0 \qquad (a \neq 0) \tag{13}$$

The differential equation   Section 3.1

$$\frac{\mathrm{d}^2y}{\mathrm{d}x^2} - \lambda^2 y = 0 \tag{10}$$

has a general solution of the form

$$y = p\exp\lambda x + q\exp(-\lambda x)$$

where p and q are arbitrary constants. Values for p and q can be found from
two initial or boundary conditions.

The differential equation   Section 3.2

$$\frac{\mathrm{d}^2y}{\mathrm{d}x^2} + \lambda^2 y = 0 \tag{9}$$

has a general solution of the form

$$y = p\sin\lambda x + q\cos\lambda x$$

where p and q are arbitrary constants. This function is called a *sinusoid*.
Again, values for p and q can be found from two initial or boundary
conditions.

The solution of differential equation (13),   Section 3.3

$$a\frac{\mathrm{d}^2y}{\mathrm{d}x^2} + b\frac{\mathrm{d}y}{\mathrm{d}x} + cy = 0 \qquad (a \neq 0)$$

depends on the relative values of a, b and c.

1   When $b^2 > 4ac$ the solution is

$$y = p\exp\left\{\frac{-b + \sqrt{(b^2 - 4ac)}}{2a}x\right\} + q\exp\left\{\frac{-b - \sqrt{(b^2 - 4ac)}}{2a}x\right\}$$

where p and q are arbitrary constants.

2   When $b^2 = 4ac$ the solution is

$$y = \exp\left(-\frac{bx}{2a}\right)(px + q)$$

where p and q are arbitrary constants.

3   When $b^2 < 4ac$ the solution is

$$y = \exp\left(-\frac{bx}{2a}\right)\left[p\sin\left\{\frac{\sqrt{(4ac - b^2)}}{2a}x\right\} + q\cos\left\{\frac{\sqrt{(4ac - b^2)}}{2a}x\right\}\right]$$

where p and q are arbitrary constants.

A differential equation of the form $\mathrm{d}^2y/\mathrm{d}x^2 - \lambda^2 y = 0$ arises out of a model   **Section 4**
of heat flow along a cooling fin and can arise out of other models of physical
situations. The model of heat flow can be used to answer questions about
the operating temperature of an engine, etc.

ANSWERS TO SELF-ASSESSMENT QUESTIONS

SAQ 1

$$Y = 100 - 4.9T^2 \qquad (2)$$

$$\frac{dY}{dT} = -2 \times 4.9T$$

$$= -9.8T$$

$$\frac{d^2Y}{dT^2} = -9.8$$

This is a differential equation (1), hence equation (2) is indeed a solution.

SAQ 2

The differential equation is

$$\frac{d^2Y}{dT^2} = -\frac{9.8}{6} = -1.6 \quad \text{(to two significant figures)}$$

Its general solution is obtained by integrating twice

$$\frac{dY}{dT} = \int -1.6 \, dT$$

$$= -1.6T + B$$

$$Y = \int (-1.6T + B) \, dT$$

$$Y = -\frac{1.6T^2}{2} + BT + C$$

$$= -0.8T^2 + BT + C$$

This is the required general solution.

SAQ 3

Let the lowest height from which he can safely jump be M metres. Then the solution in this case is

$$Y = -4.9T^2 + M$$

But when $T = 4$, Y must be at least 200 m. So I can put $Y = 200$ when $T = 4$ and hence find M.

$$200 = -4.9 \times 16 + M$$

$$M = 200 + 4.9 \times 16$$

$$= 278 \quad \text{(to three significant figures)}$$

So he should jump from at least 280 metres.

SAQ 4

The general solution (see SAQ 2) is

$$Y = -0.8T^2 + BT + C$$

The initial conditions are $Y = 500$, $dY/dT = 1500$ when $T = 0$. (Both Y and dY/dT are positive because the spaceship is *above* the Moon's surface and is travelling *upwards*.)

Putting $Y = 500$ when $T = 0$ into the general solution gives

$$500 = 0 + 0 + C$$

So $C = 500$

To use the other initial condition I shall have to differentiate the general solution:

$$\frac{dY}{dT} = -1.6T + B$$

Putting $dY/dT = 1500$ when $T = 0$ gives

$$1500 = 0 + B$$

So $B = 1500$.

Hence, the particular solution is

$$Y = -0.8T^2 + 1500T + 500$$

SAQ 5

The general solution is obtained by integrating twice:

$$\frac{dX}{dT} = \int 5 \, dT$$

$$= 5T + B$$

where B is an arbitrary constant

$$X = \int (5T + B) \, dT$$

$$= \frac{5T^2}{2} + BT + C$$

where C is a second arbitrary constant.

The particular solution is found by using the initial conditions.

Putting $X = 3$ when $T = 1$ gives

$$3 = \tfrac{5}{2} \times 1 + B \times 1 + C$$

so $\tfrac{1}{2} = B + C$

Putting $dX/dT = 1$ when $T = 1$ into the expression for dX/dT gives

$$1 = 5 \times 1 + B$$

so $B = -4$

Hence $\tfrac{1}{2} = -4 + C$

and $C = 4\tfrac{1}{2}$

The particular solution is therefore

$$X = 2\tfrac{1}{2}T^2 - 4T + 4\tfrac{1}{2}$$

SAQ 6

(a) Let the output be P million units per year. The model for each factory is

$$\frac{d^2P}{dT^2} = 0.005$$

Then the general solution of this is

$$P = 0.0025T^2 + BT + C$$

When $T = 0$, $P = 1$ and thus $C = 1$. The common equation for output against time is therefore

$$P = 0.0025T^2 + BT + 1$$

(b) The growth rate of a factory is dP/dT and therefore the general equation for growth rate is

$$\frac{dP}{dT} = 0.005T + B$$

Since B does not appear in the second-order differential equation, the model for all factories still holds even if B varies from factory to factory. If one group of factories has a slower growth rate, the initial growth rate B ($B = dP/dT$ when $T = 0$) must be lower.

(c) If the initial growth rate is zero, then $B = 0$ and

$$P = 0.0025T^2 + 1.$$

SAQ 7

The general solution is obtained by integrating twice:

$$\frac{dY}{dT} = \int 1 \, dT$$

$$= T + B$$

where B is an arbitrary constant.

$$Y = \int (T + B) \, dT$$

$$= \frac{T^2}{2} + BT + C$$

where C is a second arbitrary constant.

The particular solution is found by using the boundary conditions.

Putting $Y = 3$ when $T = 0$ gives

$$3 = 0 + 0 + C$$

so

$$C = 3$$

Putting $Y = 5$ when $T = 2$ gives

$$5 = \frac{2^2}{2} + 2B + 3$$

So $B = 0$ and the particular solution is

$$Y = \frac{T^2}{2} + 3$$

SAQ 8

The general equation is

$$Y = -4.9T^2 + BT + C$$

The boundary conditions are

$$Y = 50 \quad \text{when} \quad T = 0$$

and

$$Y = 0 \quad \text{when} \quad T = T_0$$

Substituting these values into the general solution gives

$$50 = C$$

and

$$0 = -4.9T_0^2 + BT_0 + C$$

The first substitution gives C directly and the second makes it possible to find B. Thus

$$B = \frac{4.9T_0^2 - 50}{T_0}$$

With the two integration constants evaluated you can now obtain a general equation for distance as a function of T, namely

$$Y = -4.9T^2 + (4.9T_0^2 - 50)\frac{T}{T_0} + 50$$

This gives you Y for any given value of T.

SAQ 9

For each solution it is necessary to integrate the right-hand side of each equation twice

(a) $\quad \dfrac{dY}{dT} = \int (-10 + 3T) \, dT$

$$= -10T + \frac{3}{2}T^2 + B$$

and

$$Y = 5T^2 + \frac{1}{2}T^3 + BT + C$$

where B and C are arbitrary constants.

(b) $\quad \dfrac{dY}{dT} = \int (\exp 2T + \exp T) \, dT$

$$= \frac{1}{2}\exp 2T + \exp T + B$$

and

$$Y = \frac{1}{4}\exp 2T + \exp T + BT + C$$

where B and C are arbitrary constants.

(c) $\quad \dfrac{dy}{dt} = \int c \sin \omega t$

$$= -\frac{c}{\omega}\cos \omega t + d$$

and

$$y = -\frac{c}{\omega^2}\sin \omega t + dt + e$$

where d and e are arbitrary constants.

SAQ 10

(a) The general solution is obtained by integrating twice:

$$\frac{dY}{dT} = \int 5T^2 \, dT$$

$$= \frac{5T^3}{3} + B$$

where B is an arbitrary constant.

$$Y = \int \left(\frac{5T^3}{3} + B\right) dT$$

$$= \frac{5T^4}{12} + BT + C$$

where C is a second arbitrary constant.

The particular solution is found by using the initial conditions.

Putting $Y = 3$ at $T = 0$ gives

$$3 = 0 + 0 + C$$

so $C = 3$.

Putting $dY/dT = 1$ at $T = 0$ into the expression for dY/dT gives

$$1 = 0 + B$$

So $B = 1$ and the particular solution is

$$Y = \frac{5T^4}{12} + T + 3$$

(b) The general solution is obtained by integrating twice:

$$\frac{dX}{dT} = \int 3 \cos T \, dT$$

$$= 3 \sin T + B$$

where B is an arbitrary constant.

$$X = \int (3 \sin T + B) \, dT$$

$$= -3 \cos T + BT + C$$

where C is a second arbitrary constant.

The particular solution is found by using the boundary conditions.

Putting $X = 1$ at $T = 0$ gives

$$1 = -3\cos 0 + 0 + C$$

$$= -3 + C$$

so $C = 4$.

Putting $X = 4$ at $T = \pi/2$ gives

$$4 = -3\cos \pi/2 + B\pi/2 + 4$$

$$= 0 + B\pi/4 + 4$$

So

$B = 0$ and the particular solution is

$Y = -3\cos T + 4$

SAQ 11

The general solution for this car was

$$Y = T^2 + \frac{0.5T^3}{6} + BT + C$$

where B and C were arbitrary constants. The particular solution can be found from the boundary conditions. Putting $Y = 10$ when $T = 0$ gives

$$10 = 0 + 0 + 0 + C$$

Hence $C = 10$.

Putting $Y = 100$ when $T = 6$ gives

$$100 = 36 + \frac{0.5 \times 6 \times 36}{6} + 6B + 10$$

$$= 36 + 18 + 6B + 10$$

$$36 = 6B$$

$$B = 6$$

The particular solution is therefore

$$Y = T^2 + \frac{0.5T^3}{6} + 6T + 10$$

At $T = 0$, $Y = 10$.

At $T = 3$, $Y = 39\frac{1}{4}$.

So between $T = 0$ and $T = 3$ the car travelled $29\frac{1}{4}$ m.

SAQ 12

(a) The equation is the general solution of the differential equation.

(b) The equation given can be reduced to the general solution by writing $bt + ct$ as et, where $e\,(=b+c)$ is one of the constants of integration.

(c) This equation is not the general solution. The two constants can be combined, as in (b) so that there is only one constant of integration in the solution.

SAQ 13

y is the dependent variable.

t is the independent variable.

a is a constant parameter (the acceleration).

c is a constant of integration depending on the initial value of y.

b is a constant of integration depending on the initial value of dy/dt.

SAQ 14

$$y = p \exp kx$$

$$\frac{dy}{dx} = pk \exp kx$$

$$\frac{d^2y}{dx^2} = pk^2 \exp kx$$

Taking differential equation (9) first,

$$\frac{d^2y}{dx^2} + \lambda^2 y = 0$$

I can substitute for y and for d^2y/dx^2:

$$pk^2 \exp kx + p\lambda^2 \exp kx = 0$$

$$p \exp kx(k^2 + \lambda^2) = 0$$

Now $\exp kx$ is not zero for any finite value of x, hence if this expression is to be zero it must be that

$$k^2 + \lambda^2 = 0$$

or

$$k^2 = -\lambda^2$$

Both k^2 and λ^2 must be *positive* for any real (or normal) number, hence this condition is *not* satisfied and $y = p \exp kx$ is not a solution of differential equation (9).

Taking differential equation (10)

$$\frac{d^2y}{dx^2} - \lambda^2 y = 0$$

I can substitute for y and d^2y/dx^2

$$pk^2 \exp kx - p\lambda^2 \exp kx = 0$$

$$p \exp kx(k^2 - \lambda^2) = 0$$

This expression is zero if

$$k^2 - \lambda^2 = 0$$

or

$$k^2 = \lambda^2.$$

This condition can be satisfied if either $k = \lambda$ or $k = -\lambda$, hence $y = p \exp kx$ *is* a solution of differential equation (10) provided $k = \pm\lambda$.

SAQ 15

(a) $y = p \exp \lambda x + b$

$$\frac{dy}{dx} = p\lambda \exp \lambda x$$

$$\frac{d^2y}{dx^2} = p\lambda^2 \exp \lambda x$$

Substituting into differential equation (10) gives

$$p\lambda^2 \exp \lambda x - p\lambda^2 \exp \lambda x - \lambda^2 b = 0$$

This expression is not zero unless b is zero, which is forbidden. Hence the equation in (a) is not a solution.

(b) $y = p \exp(-\lambda x) + cx$

$$\frac{dy}{dx} = -\lambda p \exp(-\lambda x) + c$$

$$\frac{d^2y}{dx^2} = \lambda^2 p \exp(-\lambda x)$$

Substituting into differential equation (10) gives

$$\lambda^2 p \exp(-\lambda x) - \lambda^2 p \exp(-\lambda x) - \lambda^2 cx = 0$$

This expression is not zero for every value of x unless c is zero, which is forbidden. Hence, the equation in (b) is not a solution.

(c) $y = p \exp \lambda x + q \exp(-\lambda x)$

$$\frac{dy}{dx} = p\lambda \exp \lambda x - q\lambda \exp(-\lambda x)$$

$$\frac{d^2y}{dx^2} = p\lambda^2 \exp \lambda x + q\lambda^2 \exp(-\lambda x)$$

Substituting into differential equation (10) gives

$$p\lambda^2 \exp \lambda x + q\lambda^2 \exp(-\lambda x) - \lambda^2 p \exp \lambda x - \lambda^2 q \exp(-\lambda x) = 0$$

This equality holds whatever the values of p and q, hence the equation in (c) is a solution.

SAQ 16

In all the solutions, P and Q are arbitrary constants

(a)   Here 4 corresponds to λ^2. Hence, the general solution is

$$Y = P \exp 2X + Q \exp(-2X)$$

(b)   Here 16 corresponds to λ^2. Hence, the general solution is

$$Y = P \exp 4X + Q \exp(-4X)$$

(c)   Here 9 corresponds to λ^2. Hence, the general solution is

$$Y = P \exp 3X + Q \exp(-3X)$$

SAQ 17

(a) The general solution is

$$Y = P \exp T + Q \exp(-T)$$

where P and Q are arbitrary constants. Putting $Y = 6$ at $T = 0$,

$$6 = P \exp 0 + Q \exp 0$$
$$= P + Q$$

To use the other initial condition the expression for Y must be differentiated:

$$\frac{dY}{dT} = P \exp T - Q \exp(-T)$$

Putting $dY/dT = 0$ at $T = 0$ gives

$$0 = P - Q$$

Hence $P = Q$ and so $P = Q = 3$. The particular solution is

$$Y = 3 \exp T + 3 \exp(-T)$$

(b) The general solution is

$$X = P \exp 2T + Q \exp(-2T)$$

where P and Q are arbitrary constants. Putting $X = 2$ at $T = 0$,

$$2 = P \exp 0 + Q \exp 0$$
$$= P + Q$$

Putting $X = 4$ at $T = 1$,

$$4 = P \exp 2 + Q \exp(-2)$$
$$= Pe^2 + Qe^{-2}$$

Multiply through by e^2:

$$4e^2 = Pe^4 + Q$$

Also

$$2 = P + Q$$

Subtract $4e^2 - 2 = P(e^4 - 1)$

So

$$P = \frac{4e^2 - 2}{e^4 - 1}$$

Also

$$Q = 2 - P$$

So $Q = 2 - \dfrac{4e^2 - 2}{e^4 - 1}$

$$= \frac{2e^4 - 2 - 4e^2 + 2}{e^4 - 1}$$

$$= \frac{2e^4 - 4e^2}{e^4 - 1}$$

Hence the particular solution is

$$X = \frac{4e^2 - 2}{e^4 - 1} e^{2T} + \frac{2e^4 - 4e^2}{e^4 - 1} e^{-2T}$$

SAQ 18

(a) $y = p \sin kx$

$$\frac{dy}{dx} = pk \cos kx$$

$$\frac{d^2y}{dx^2} = -pk^2 \sin kx$$

Substituting into equation (9) gives

$$-pk^2 \sin kx + \lambda^2 p \sin kx = 0$$
$$p \sin kx(\lambda^2 - k^2) = 0$$

This equation can be satisfied for *any* value of x if $\lambda^2 = k^2$, that is $k = \pm\lambda$. Hence, the given expression is a solution provided $k = \pm\lambda$.

(b) $y = p \cos kx$

$$\frac{dy}{dx} = -pk \sin kx$$

$$\frac{d^2y}{dx^2} = -pk^2 \cos kx$$

Substituting into equation (9) gives

$$-pk^2 \cos kx + \lambda^2 p \cos kx = 0$$
$$p \cos kx(\lambda^2 - k^2) = 0$$

Once again, this equation is satisfied for any value of x if $k = \pm\lambda$. Hence, the given expression is a solution provided $k = \pm\lambda$.

SAQ 19

(a) P and Q or p and q represent arbitrary constants

(i) Here 25 corresponds to λ^2. So the general solution is

$$Y = P \sin 5X + Q \cos 5X$$

(ii) Rearrange the equation:

$$\frac{d^2Y}{dT^2} + \frac{4}{9} Y = 0$$

Here $\frac{4}{9}$ corresponds to λ^2. So the general solution is

$$Y = P \sin\left(\frac{2T}{3}\right) + Q \cos\left(\frac{2T}{3}\right)$$

(iii) Here ω^2 corresponds to λ^2. So the general solution is
$$y = p \sin \omega t + q \cos \omega t$$

(b) Here the general solution is

$$Y = P \sin 4X + Q \cos 4X$$

where P and Q are arbitrary constants. Putting $Y = 0$ at $X = 0$ gives

$$0 = P \sin 0 + Q \cos 0$$

$$= 0 + Q$$

So $Q = 0$

Putting $Y = 3$ at $X = \pi/8$ gives

$$3 = P \sin \pi/2 + Q \cos \pi/2$$

$$= P + 0$$

Hence $P = 3$ and the particular solution is $Y = 3 \sin 4X$

SAQ 20

(a) This is like differential equation (9) and 1 corresponds to λ^2. The general solution is therefore

$$Y = P \sin X + Q \cos X$$

where P and Q are arbitrary constants. Putting $Y = 0$ at $X = 0$ gives

$$0 = P \sin 0 + Q \cos 0$$

$$= 0 + Q$$

Hence $Q = 0$.

To use the other condition, the expression for Y must be differentiated:

$$\frac{dY}{dX} = P \cos X + 0 \quad \text{(because } Q = 0\text{)}$$

Putting $dY/dX = 3$ at $X = 0$ gives

$$3 = P \cos 0$$

So $P = 3$ and the particular solution is

$$Y = 3 \sin X$$

(b) This is like differential equation (10) and 4 corresponds to λ^2. The general solution is therefore

$$Y = P \exp 2T + Q \exp(-2T)$$

where P and Q are arbitrary constants. Putting $Y = 0$ at $T = 0$ gives

$$0 = P \exp 0 + Q \exp 0$$

$$0 = P + Q$$

To use the other condition the expression for Y must be differentiated:

$$\frac{dY}{dT} = 2P \exp 2T - 2Q \exp(-2T)$$

Putting $dY/dT = 4$ at $T = 0$ gives

$$4 = 2P \exp 0 - 2Q \exp 0$$

$$= 2P - 2Q$$

so $\quad 2 = P - Q$

and $\quad 0 = P + Q$.

Adding $\quad 2 = 2P$,

So $P = 1$ and $Q = -1$. Hence, the particular solution is

$$Y = \exp 2T - \exp(-2T)$$

SAQ 21

$y = p \exp kx$

$$\frac{dy}{dx} = pk \exp kx$$

$$\frac{d^2y}{dx^2} = pk^2 \exp kx$$

Substituting into differential equation (13) gives

$$apk^2 \exp kx + bpk \exp kx + cp \exp kx = 0$$

$$p \exp kx(ak^2 + bk + c) = 0$$

This expression equals zero if $ak^2 + bk + c = 0$.

This is a quadratic equation for k with solution

$$k = \frac{-b \pm \sqrt{(b^2 - 4ac)}}{2a}$$

Provided k has one of these two values, $y = p \exp kx$ is a solution of differential equation (13).

SAQ 22

(a) Check $B^2 > 4AC$:

$$B^2 = 9$$

$$4AC = 8$$

So $B^2 > 4AC$ and the general solution is

$$Y = P \exp \left(\frac{-3 + \sqrt{(9 - 8)}}{2} X \right)$$

$$+ Q \exp \left(\frac{-3 - \sqrt{(9 - 8)}}{2} X \right)$$

where P and Q are arbitrary constants.

This simplifies to

$$Y = P \exp(-X) + Q \exp(-2X)$$

(b) Check $B^2 > 4AC$:

$$B^2 = 289$$

$$4AC = 288$$

So $B^2 > 4AC$ and the general solution is

$$Y = P \exp \left(\frac{-17 + \sqrt{(289 - 288)}}{12} T \right)$$

$$+ Q \exp \left(\frac{-17 - \sqrt{(289 - 288)}}{12} T \right)$$

where P and Q are arbitrary constants.

This simplifies to

$$Y = P \exp \left(-\frac{4T}{3} \right) + Q \exp \left(-\frac{3T}{2} \right)$$

SAQ 23

(a) $y = p \exp kx \sin mx$

Use the product rule with

$$g = p \exp kx$$

$$h = \sin mx$$

Then

$$\frac{dg}{dx} = pk \exp kx$$

$$\frac{dh}{dx} = m \cos mx$$

$$\frac{dy}{dx} = g\frac{dh}{dx} + \frac{dg}{dx}h$$

$$= p \exp kx \, m \cos mx + pk \exp kx \sin mx$$

$$= p \exp kx(m \cos mx + k \sin mx)$$

(b) Use the product rule again with

$$g = p \exp kx$$

$$h = m \cos mx + k \sin mx$$

After tidying up, this yields the given expression.

(c) Substitution gives

$$ap \exp kx\{(k^2 - m^2)\sin mx + 2km \cos mx\}$$

$$+ bp \exp kx(m \cos mx + k \sin mx)$$

$$+ cp \exp kx \sin mx = 0$$

$p \exp kx$ is a common factor and the sine and cosine terms can be tidied up to give

$$p \exp kx\{(ak^2 - am^2 + bk + c)\sin mx$$

$$+ (2akm + bm)\cos mx\} = 0$$

(d) Setting the coefficient of $\sin mx$ equal to zero gives

$$ak^2 - am^2 + bk + c = 0$$

Setting the coefficient of $\cos mx$ equal to zero gives

$$2akm + bm = 0$$

From this latter equation, either $m = 0$ or $k = -b/2a$. Because $m = 0$ would not be a useful solution, it must be that $k = -b/2a$. Using this in the first equation gives

$$\frac{ab^2}{4a^2} - am^2 - \frac{b^2}{2a} + c = 0$$

Hence

$$am^2 = \frac{ab^2}{4a^2} - \frac{b^2}{2a} + c$$

$$= \frac{ab^2 - 2ab^2 + 4a^2c}{4a^2}$$

$$= \frac{4a^2c - ab^2}{4a^2}$$

Hence

$$m^2 = \frac{4ac - b^2}{4a^2}$$

and

$$m = \frac{\pm\sqrt{(4ac - b^2)}}{2a}$$

These values of k and m are the required ones.

SAQ 24

(a) Check $B^2 < 4AC$:

$$B^2 = 4$$

$$4AC = 8$$

So $B^2 < 4AC$ and the general solution is

$$Y = \exp\left(-\frac{2X}{2}\right)\left[P \sin\left\{\frac{\sqrt{(8 - 4)}}{2}X\right\}\right.$$

$$\left. + Q \cos\left\{\frac{\sqrt{(8 - 4)}}{2}X\right\}\right]$$

where P and Q are arbitrary constants. This simplifies to
$Y = \exp(-X)(P \sin X + Q \cos X)$

(b) Check $B^2 < 4AC$

$$B^2 = 16$$

$$4AC = 20$$

So $B^2 < 4AC$ and the general solution is

$$Y = \exp\left(-\frac{4X}{2}\right)\left[P \sin\left\{\frac{\sqrt{(20 - 16)}}{2}X\right\}\right.$$

$$\left. + Q \cos\left\{\frac{\sqrt{(20 - 16)}}{2}X\right\}\right]$$

where P and Q are arbitrary constants. This simplifies to
$Y = \exp(-2X)(P \sin X + Q \cos X)$

SAQ 25

(a) In this case $B^2 = 36$ and $4AC = 32$ so $B^2 > 4AC$. The general solution is therefore

$$Y = P \exp\left(\frac{-6 + \sqrt{(36 - 32)}}{2}T\right)$$

$$+ Q \exp\left(\frac{-6 - \sqrt{(36 - 32)}}{2}T\right)$$

where P and Q are arbitrary constants. This simplifies to
$Y = P \exp(-2T) + Q \exp(-4T)$

(b) In this case $B^2 = 36$ and $4AC = 40$ so $B^2 < 4AC$. The general solution is therefore

$$X = \exp\left(-\frac{6T}{2}\right)\left[P \sin\left\{\frac{\sqrt{(40 - 36)}}{2}T\right\}\right.$$

$$\left. + Q \cos\left\{\frac{\sqrt{(40 - 36)}}{2}T\right\}\right]$$

where P and Q are arbitrary constants. This simplifies to
$X = \exp(-3T)(P \sin T + Q \cos T)$

SAQ 26

The differential equation

$$\frac{d^2y}{dx^2} - \lambda^2 y = 0 \text{ has the general solution}$$

$$y = p \exp \lambda x + q \exp(-\lambda x)$$

where p and q are arbitrary constants. By analogy, differential equation (32) has the general solution

$$q = p \exp\left\{\sqrt{\left(\frac{2j(b + w)}{\kappa b w}\right)}x\right\}$$

$$+ m \exp\left\{-\sqrt{\left(\frac{2j(b + w)}{\kappa b w}\right)}x\right\}$$

where I have used m instead of q as one of the arbitrary constants, because q is already being used in the model.

SAQ 27

To answer this you need to go through the process described in the text. First, differentiate equation (27),

$$\frac{dQ}{dX} = -16.2 \times 15.8 \exp 15.8X - 78.7 \times 15.8 \exp(-15.8X)$$

next use the fact that

$$\frac{dQ}{dX} = -2J(W + B)\theta$$

So

$$\theta = \frac{-1}{2J(W + B)}$$

$$- 78.7 \times 15.8 \exp(-15.8X)\}$$

The operating temperature is the value of θ at the hot end of the fin—that is, θ at $X = 0$. I shall call this value θ_0.

$$\theta_0 = \frac{-1}{2J(W + B)}(-16.2 \times 15.8 \exp 0 - 78.7 \times 15.8 \exp 0)$$

Putting in $\exp 0 = 1$, $J = 25$, $W = 0.2$ and $B = 0.001$ gives

$$\theta_0 = \frac{1}{2 \times 25 \times (0.2 + 0.001)}(16.2 \times 15.8 + 78.7 \times 15.8)$$

$$= \frac{256 + 1243}{10.05}$$

$$= 149 \text{ (to three significant figures)}$$

This is below the 200 K maximum excess temperature and so the modification is probably safe.

SAQ 28

λ is given by

$$\lambda = \sqrt{\left\{\frac{2j(b + w)}{\kappa b w}\right\}}$$

and so its numerical value is

$$\sqrt{\left(\frac{2 \times 5 \times 0.101}{385 \times 0.1 \times 0.001}\right)}$$

$$= \sqrt{26.2}$$

$$= 5.12 \text{ (to three significant figures)}$$

$\theta_0 = 100\,\text{K}$ and $l = 0.1\,\text{m}$.

Hence equation (32) is, in numerical form,

$$\theta = \frac{100\{\exp 5.12X + \exp(1.024 - 5.12X)\}}{1 + \exp 1.024}$$

To evaluate the temperature of the remote edge of the fin means finding θ when $X = 0.1$. Hence

$$\theta = \frac{100\{\exp 0.512 + \exp(1.024 - 0.512)\}}{1 + \exp 1.024}$$

$$= \frac{200 \exp 0.512}{1 + \exp 1.024}$$

$$= \frac{200 \times 1.669}{1 + 2.784}$$

$$= 88 \text{ (to two significant figures)}$$

The temperature at the tip of the fin is therefore approximately 88 K above the air temperature.

15. Growth, Decay and Oscillation 2

CONTENTS

AIMS

The aims of this unit are:

1   To show that a number of periodic phenomena may be modelled by a differential equation of the form $\ddot{y} + \omega^2 y = 0$ and to investigate the nature of the solution of this differential equation.

2   To show that this model may be too simple for some purposes because it neglects any resistive force; to set up the differential equation which includes such a force and to investigate the nature of the solution of this differential equation.

OBJECTIVES

After reading this unit you should be able to:

1   Distinguish between true and false statements concerning, or explain in your own words the meaning of, the following terms:

amplitude   frequency
angular frequency   periodic time or period
damped oscillation   phase
damped simple harmonic motion simple harmonic motion
decaying oscillation   spring constant

2   Convert an expression in the form $y = p \sin \omega t + 1q \cos \omega t$ into an expression in the form $y = r \sin(\omega t + \varepsilon)$. (SAQ 5)

3   Calculate or state the periodic time, frequency, angular frequency, amplitude and phase for a particular simple harmonic motion and the periodic time, frequency and angular frequency for a particular damped simple harmonic motion. (SAQs 7–15)

4   Use differential equations of the type introduced in Section 3 of Unit 14 in particular modelling situations, and find general and/or particular solutions of them. (SAQs 13, 14 and 17–19)

5   Recognize whether a particular model will lead to simple harmonic motion. (SAQ 16)

6   Describe and sketch the effects of damping on a system which, in the absence of damping, would execute simple harmonic motion. (SAQs 18–20)

STUDY GUIDE

Your work for this study week consists of Unit 15 *Growth, Decay and Oscillation 2* and TV9 *Braking for Safety*.

This unit follows on directly from Unit 14. It looks at how some of the differential equations introduced in that unit are used in models. The unit also builds to some extent on material first introduced in Units 4 and 9. You are strongly advised to attempt the first four SAQs, which are revision SAQs, and to refer back to the appropriate sections of Units 4, 9 or 14 if you have any difficulty in answering them.

TV9 *Braking for Safety* is related to Section 4, and you will find the programme more useful if you have read that section, although this is not essential.

As pointed out in the study guide to Unit 14, that unit is longer than Unit 15. You should plan your study of Unit 15 with this in mind.

Details of any assignments associated with Unit 15 are given in the *Supplementary Material* mailed with Units 14–16.

4

1 INTRODUCTION

In this unit you will continue the study of second-order differential equations. In Unit 14 you met two types of second-order differential equation. They either took the form $d^2y/dx^2 = f(x)$, where both $f(x)$ and the integral of $f(x)$ can be integrated, or $a\,d^2y/dx^2 + b\,dy/dx + cy = 0$, where a, b and c are constants and $a \neq 0$. Unit 15 will deal with certain members of the latter family of differential equations. It will look at mathematical models which give rise to differential equations of this type and at the significance of the solutions of these equations.

Before introducing the main topic of this unit, which is the modelling of oscillations, I shall first deal with some mathematical preliminaries.

1.1 Revision and preliminaries

It will help you in your study of this unit if you first try the four revision SAQs below, and check back to the appropriate course material if you have any difficulties.

After these SAQs I shall discuss an alternative form for the solution of one of the types of differential equation you met in Unit 14

$$\frac{d^2y}{dx^2} + \lambda^2 y = 0$$

This alternative form is necessary for some of the subsequent work in the unit.

SAQ 1 (Revision: Unit 4, Section 6.4 and Unit 9, Section 1.2)   SAQ 1

(a) Sketch the graph of $Y = \sin X$ for $-2\pi \leqslant X \leqslant 2\pi$.

Write down the values of X for which

(i) $Y = 1$

(ii) $Y = 0$

(iii) $Y = -1$

(b) Sketch the graph of $Y = 3\cos X$ for $-2\pi \leqslant X \leqslant 2\pi$.

Write down the values of X for which

(i) $Y = 3$

(ii) $Y = 0$

(iii) $Y = -3$

(c) Sketch the graph of $Y = \cos 4X$ for $-2\pi \leqslant X \leqslant 2\pi$.

Write down the values of X for which $Y = 0$.

SAQ 2 (Revision: Unit 14, mainly Section 3.2)   SAQ 2

(a) How many arbitrary constants would you expect in the general solution of a second-order differential equation?

(b) What is the general solution of

$$\frac{d^2Y}{dX^2} + 9Y = 0$$

(c) If $Y = 0$ and $dY/dX = 3$ at $X = 0$, find the particular solution for this case.

SAQ 3 (Revision: Unit 14, Section 3.3)

(a) Under what conditions will the differential equation

$$a\frac{d^2y}{dx^2} + b\frac{dy}{dx} + cy = 0 \qquad (a \neq 0)$$

have

(i) a solution which is an exponential times a sinusoid;

(ii) a solution which is a sum or difference of two exponentials?

(b) Find the general solution of

$$\frac{d^2Y}{dT^2} + 2\frac{dY}{dT} + 5Y = 0$$

SAQ 4 (Revision: Unit 9, Section 2.2)

(a) If

$$y = r\sin(\omega t + \varepsilon)$$

where r, ω and ε are constants, find $\dot{y}$ and $\ddot{y}$.

(Remember that $\dot{y}$ means dy/dt and $\ddot{y}$ means d^2y/dt^2.)

(b) Is the expression $y = r\sin(\omega t + \varepsilon)$ a solution of the differential equation

$$\ddot{y} + \omega^2 y = 0?$$

Your answer to part (b) of SAQ 4 may have surprised you. From Unit 14, you would expect the solution of the differential equation

$$\frac{d^2y}{dt^2} + \omega^2 y = 0$$

to be

$$y = p\sin\omega t + q\cos\omega t \tag{1}$$

where p and q are arbitrary constants. So why should

$$y = r\sin(\omega t + \varepsilon) \tag{2}$$

also turn out to be a solution; and, moreover, a solution with two arbitrary constants, r and ε?

At first sight it might seem that there are two entirely different forms of general solution. In fact, there are simply two different forms of the *same* general solution. To show you that this is the case I shall have to use one of the relationships among the trigonometric ratios which is given in your *Handbook*.

Study comment

You would not be expected to reproduce any of the following arguments, as far as SAQ 5. You would be expected to use the results, equations (3) to (6), as illustrated in the examples.

The expression I shall use is

$$\sin(\alpha + \beta) = \sin\alpha\cos\beta + \sin\beta\cos\alpha$$

I shall apply this expression to equation (2):

$$y = r\sin(\omega t + \varepsilon)$$

$$= r[\sin\omega t\cos\varepsilon + \sin\varepsilon\cos\omega t]$$

$$= r\cos\varepsilon\sin\omega t + r\sin\varepsilon\cos\omega t$$

Now $r\cos\varepsilon$ and $r\sin\varepsilon$ are constants because both r and ε are constants. Therefore, in this expanded form I have an expression very close to equation (1),

$$y = p\sin\omega t + q\cos\omega t$$

In fact, the two expressions are identical if

$$r\cos\varepsilon = p \tag{3}$$

$$r\sin\varepsilon = q \tag{4}$$

Therefore, equations (1) and (2) are two alternative ways of writing the general solution of the differential equation *provided* the two arbitrary constants p and q are related to the two arbitrary constants r and ε by equations (3) and (4) above.

In this unit you need to be familiar with both these forms of general solution. On occasions, also, you will need to convert a particular solution from one form to the other.

To convert from a particular solution found from the general solution $y = r\sin(\omega t + \varepsilon)$, you simply give values to p and q using equations (3) and (4) above.

Example

A particular solution of the differential equation $\ddot{Y} + 9Y = 0$ has been found to be $Y = 6\sin(3T + \pi/3)$

Write this solution in the form $Y = P\sin 3T + Q\cos 3T$.

In this particular solution, $R = 6$ and $\varepsilon = \pi/3$. From equation (3)

$$P = 6\cos\left(\frac{\pi}{3}\right)$$

$$= 6 \times \tfrac{1}{2}$$

$$= 3$$

From equation (4)

$$Q = 6\sin\left(\frac{\pi}{3}\right)$$

$$= \frac{6\sqrt{3}}{2}$$

$$= 3\sqrt{3}$$

$$= 5.2 \qquad \text{(to two significant figures)}$$

Therefore, the solution in the alternative form is

$$Y = 3\sin 3T + 5.2\cos 3T$$

To convert from a particular solution found from the general solution $y = p\sin\omega t + q\cos\omega t$, it is necessary to rearrange equations (3) and (4)

$$r\cos\varepsilon = p \tag{3}$$

$$r\sin\varepsilon = q \tag{4}$$

First I square each equation and then add the two resulting equations

$$r^2\cos^2\varepsilon + r^2\sin^2\varepsilon = p^2 + q^2$$

so

$$r^2(\cos^2\varepsilon + \sin^2\varepsilon) = p^2 + q^2$$

You learned in Unit 4 that $\sin^2 \beta + \cos^2 \beta = 1$, whatever the value of β. (This result is also quoted in the *Handbook*.) Hence

$$r^2 = p^2 + q^2$$

Conventionally, r is taken to be positive. Therefore

$$r = \sqrt{(p^2 + q^2)} \qquad (5)$$

This equation enables r to be found when p and q are known. To find ε I must divide equation (3) into equation (4):

$$\frac{r \sin \varepsilon}{r \cos \varepsilon} = \frac{q}{p}$$

Because $\sin \varepsilon / \cos \varepsilon = \tan \varepsilon$,

$$\tan \varepsilon = \frac{q}{p} \qquad (6)$$

This equation enables ε to be found, but not uniquely. As you will remember from Unit 4, many values of ε can all have the same value for $\tan \varepsilon$. (For instance, $\tan \varepsilon = 1$ when $\varepsilon = -3\pi/4, \pi/4, 5\pi/4$, and so on.) You will see how an appropriate value of ε is chosen in the following example. Notice that the value of ε is conventionally in radians, not degrees.

Example

A particular solution of the differential equation $\ddot{Y} + 4Y = 0$ has been found to be

$$Y = 3 \sin 2T + 4 \cos 2T$$

Write this solution in the form

$$Y = R \sin(2T + \varepsilon)$$

In this particular solution, $P = 3$ and $Q = 4$. Using equation (5) gives

$$R = \sqrt{(3^2 + 4^2)}$$
$$= 5$$

Using equation (6) gives

$$\tan \varepsilon = \tfrac{4}{3}$$
$$= 1.33$$

So $\varepsilon = 0.93$ or $\pi + (0.93)$ (to two significant figures)

These two values of ε are the only possible values in the range $0 \leqslant \varepsilon < 2\pi$. I shall not consider values outside that range. But which value should I choose? Well, P and Q are both positive, and $P = R \cos \varepsilon$ and $Q = R \sin \varepsilon$. R is positive so both $\sin \varepsilon$ and $\cos \varepsilon$ must be positive. If $\varepsilon = 0.93$ this is true; if $\varepsilon = \pi + 0.93$ then $\sin \varepsilon$ and $\cos \varepsilon$ are negative. So $\varepsilon = 0.93$ is the value I want and the alternative form of solution is

$$Y = 5 \sin(2T + 0.93)$$

SAQ 5

Write the following particular solution of the differential equation $\ddot{Y} + 16Y = 0$ in its alternative form:

$$Y = 5 \sin 4T + 12 \cos 4T$$

In general, it is easier to find a particular solution from the form $y = p \sin \omega t + q \cos \omega t$, as you have been doing in Unit 14. However, the alternative form I have just introduced is useful in describing some aspects of certain kinds of motion, as you will see later in this unit. You will be expected to recognize a solution in either form and to be able to convert from one form to the other.

Now that I have made this important point about the solutions of differential equations of the form $\ddot{y} + \omega^2 y = 0$, I shall conclude this introductory section by discussing briefly the rest of the unit.

1.2 Modelling oscillations

The main concern of the unit is the modelling of motion which repeats itself periodically—that is, *oscillation*. I shall discuss two examples of oscillatory motion; one is the up and down oscillations of a mass on the end of a spring, the other is the side to side oscillations of a pendulum. The particular example of a mass on a spring which I shall discuss first is a device called a *spring balance*. Figure 1 shows a spring balance. It is used to weigh objects. The principle of its operation is quite simple. The scalepan is suspended from a spring and when a weight is added to the scalepan the spring stretches and a pointer attached to the spring moves over a scale. The scale is calibrated using known weights, so the unknown weight of the object can be read off.

> You may wonder how oscillations come into the discussion of such a device. Can you see under what circumstances the scalepan and weight might start to oscillate?

If a weight is placed clumsily on the scalepan it will tend to oscillate for a while. Even if the scalepan is steadied when the weight is placed on it small oscillations can occur. I shall model such oscillations to determine what affects their size and frequency, and to investigate whether there is any way of eliminating them. Such information is useful in an examination of the oscillations of springs in other situations—the shock absorbers of a car, or a device for measuring acceleration which contains a spring. The model has much wider applications beyond the spring balance example, and is used again for one such situation at the end of the unit.

My other major example of oscillation is the motion of a pendulum. You will no doubt be familiar with such a device, perhaps through having watched one in action in a grandfather clock. A pendulum is simply a weight (usually called a *bob*) which is suspended from a fixed position by a rod or a string. Such a device can, of course, make more than one type of movement. I shall model only one of these types of motion, that in which the bob moves in an arc through its rest position; Figure 2(a). This side-to-side oscillation is, of course, the sort of motion a pendulum makes in a

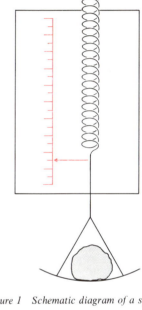

Figure 1   Schematic diagram of a spring balance

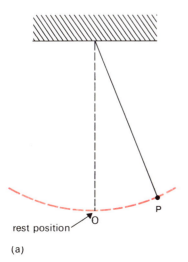

(a)

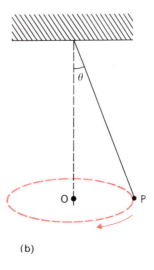

(b)

Figure 2

9

grandfather clock. I shall not be concerned with the sort of motion shown in Figure 2(b), where the pendulum moves in a circle around its rest position. This motion can be modelled using the techniques of Unit 9. The side-to-side oscillations are basic to the operation of a clock and so my model will aim to determine the length of pendulum needed for a particular clock and to examine whether there is any way of stopping the tendency of the oscillations to die away.

So in my two examples I have one case of unwanted oscillations and one of desirable oscillations. Both conditions are met in engineering, where it may be necessary to remove the one and sustain the other. To learn how to do this it is first necessary to learn something about the nature of the oscillations themselves. You will see, in the mathematical models which follow, how to predict the way in which the oscillations may be reduced or removed by making certain physical adjustments to an oscillating system. Conversely, by knowing why oscillations tend to die away, it may be possible to make physical adjustments to discourage this tendency.

So Unit 15 will look at equations which model oscillation. These equations will be derived using Newton's second law of motion, which you met in Unit 6, and will be second-order differential equations of the sort you met in Unit 14.

2 MODELLING THE OSCILLATIONS OF A SPRING BALANCE

2.1 A differential equation for the model

As indicated in Section 1, my purpose in modelling the oscillations of a spring balance is to investigate what determines whether they are large or small, fast or slow. Information about the motion might lead me to make predictions about how to eliminate the oscillations, which are a nuisance. Two of the factors which determine the nature of the oscillations are the magnitude of the mass hung on the spring and the change in the tension of the spring as it gets longer and shorter. It is also likely that the way in which the scalepan begins its oscillation may affect its subsequent behaviour. Other factors are whether the support at the upper end of the spring is truly fixed or gives a little as the scalepan moves and whether the scalepan moves sideways as well as up and down.

As a first model, I shall suppose that the motion is purely vertical and that the support is rigid. I shall also ignore any air resistance as the scalepan moves up and down or any resistance to the motion caused by the change in shape of the spring as its length changes.

Let me first look at the forces acting on the scalepan and its contents.

> What are these forces?

There are two forces. One of them is the upward pull of the spring—the *tension* in the spring. The other is the weight of the scalepan and its contents. These two forces act in opposite directions, as shown in Figure 3. If the scalepan is at rest then the upward and downward forces on it must balance each other: the upward force of the tension exactly equals the downward force of the weight.

If the scalepan moves, however, one of the forces must be larger than the other. The resultant of the two forces is the net force on the scalepan, and this net force is the one I shall be concerned with. Some of its characteristics can be readily deduced. If the scalepan is at the bottom of its 'bounce', the spring is stretched a long way and the tension is large. When the scalepan is at the top of its 'bounce' and the spring is only stretched a short way, the tension is small. Thus, the tension in the spring varies as the scalepan moves up and down. In contrast, the weight of the scalepan and its contents is constant. Thus, the *net* force varies as the scalepan moves up and down. I shall model the net force by supposing its magnitude to be proportional to the magnitude of the *displacement* of the scalepan from its position of balance, which I shall call its *rest*, or *equilibrium*, *position*. There is experimental evidence for this supposition. Figure 4 shows a sketch, based on experimental data, of how the net force and the displacement are related. Providing the displacement is not too large, the net force can be accurately modelled as being proportional to the displacement.

I also need to look at the direction in which the net force acts. Figure 4 shows me this, but let me check if it accords with common sense.

> When the scalepan is below its rest position which is larger—the tension or the weight?

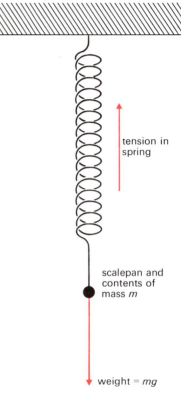

Figure 3

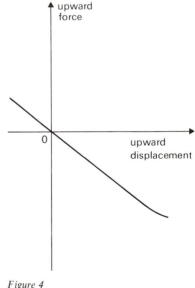

Figure 4

Remember that at the rest position the two forces are equal and that the tension increases as the spring is extended. *Below* the rest position, therefore, the tension is the larger force and the net force is *upwards*. Similarly, *above* the rest position the weight is the larger force and the net force is *downwards*. So Figure 4 does indeed accord with what we might expect. I can summarize this in one of two alternative ways:

1   the net force always acts towards the rest position; or

2   the net force always acts in the opposite direction to the displacement.

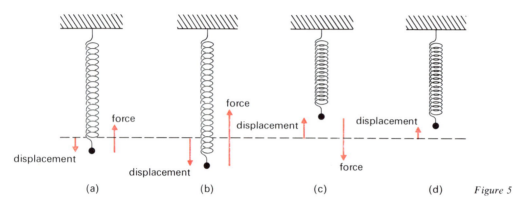

(a)   (b)   (c)   (d)   *Figure 5*

Figure 5 shows the situation. In Figure 5(a) the scalepan (represented by the black dot) is below the rest position. Its displacement is downwards and the net force on it is upwards. In Figure 5(b) the displacement is twice as large as it was in Figure 5(a) and the net force on it is twice as large also. In Figure 5(c) the scalepan is above its rest position and so the displacement is upwards and the net force downwards.

SAQ 6

Complete Figure 5(d) by drawing an arrow of appropriate length and direction to show the net force on the scalepan.

SAQ 6

I now need to express the relationship between the net force and the displacement in mathematical symbols. Because the force is proportional to the displacement but in the opposite direction I can write

$$f \propto -y$$

where f is the net force when there is a displacement y. The minus sign indicates that the force is in the opposite direction to the displacement. I shall take positive values of y or f to refer to displacements or net forces *upwards*.

Putting a constant of proportionality into this relationship gives

$$f = -ky \tag{7}$$

where k is a positive constant. The parameter k is usually called the *spring constant*. (The gradient of the straight part of the curve in Figure 4 gives a numerical value for k.) Its value is characteristic of the particular spring being used. A large value of k means that a large force is needed to produce even quite a small displacement; it denotes a stiff spring. A smaller value of k means less force is required to obtain a given displacement and denotes a less stiff spring.

spring constant

As you know from Unit 6, Newton's second law states that a net force acting on an object will cause that object to accelerate. Using the convention that dots denote derivatives with respect to time, I can denote the acceleration of the scalepan by $\ddot{y}$. Then Newton's second law gives

$$f = m\ddot{y} \tag{8}$$

Equating the two expressions for f in equations (7) and (8) gives

$$m\ddot{y} = -ky$$

This is a second-order differential equation of a type you know how to solve, and will be more apparent if I rewrite it as

$$\ddot{y} + \frac{k}{m}y = 0 \tag{9}$$

To which of the following differential equations from Unit 14 is it analogous? (Remember that k and m are positive constants.)

(a) $d^2y/dx^2 = f(x)$

(b) $d^2y/dx^2 + \lambda^2 y = 0$

(c) $d^2y/dx^2 - \lambda^2 y = 0$

It is analagous to (b), with $\lambda^2 = k/m$.

You met the solution of this sort of differential equation in Section 3.2 of Unit 14 and revised it in the Introduction to this unit. The general solution is

$$y = p\sin\left(\sqrt{\frac{k}{m}}t\right) + q\cos\left(\sqrt{\frac{k}{m}}t\right) \tag{10}$$

where p and q are arbitrary constants. As you have just seen, the general solution can also be written

$$y = r\sin\left(\sqrt{\frac{k}{m}}t + \varepsilon\right) \tag{11}$$

where r and ε are arbitrary constants which are related to p and q. In either form of solution, the important point to notice is that the solution of differential equation (9) is a sinusoid.

You will probably remember from Unit 4, or SAQ 1 of this unit, that sinusoidal curves are repetitive. I therefore have a model which predicts repetitive motion. This is to be expected in a model of the repetitive up and down motion of an object on the end of a spring. Notice that I have arrived at a solution which predicts repetitive motion and yet I have not taken into consideration the way in which the motion started. This means that oscillations are to be expected *no matter how* the motion started.

Nevertheless, I would expect the way the motion started to affect the subsequent oscillations in some way; for instance, if I pulled the scalepan down a long way and then released it I would expect larger oscillations than if I pulled it down a little way before releasing it.

Can you see how the starting conditions can be incorporated into the model?

The starting conditions lead to initial conditions—they give the values of the displacement and the velocity of the scalepan (of y and $\dot{y}$) when $t = 0$. Because the initial conditions determine the arbitrary constants p and q in equation (10) (or the arbitrary constants r and ε in equation (11)), information about the way the oscillation started is incorporated in the model when a particular solution is obtained.

As an example, I shall consider the case where the scalepan is pushed down a distance h below its rest position. One initial condition is therefore $y = -h$ at $t = 0$.

Because *upward* displacements are being considered positive and this one is downwards.

If the hand is simply removed when the contents are in place there will be no initial velocity, and $\dot{y} = 0$ at $t = 0$ is the other condition.

As I indicated in the Introduction, it is generally easier to work with the general solution in the form of equation (10),

$$y = p \sin\left(\sqrt{\frac{k}{m}}\,t\right) + q \cos\left(\sqrt{\frac{k}{m}}\,t\right) \tag{10}$$

and I shall use that form here.

Putting $y = -h$ at $t = 0$ into equation (10) gives

$$-h = p \sin 0 + q \cos 0$$

$$= 0 + q$$

hence $q = -h$.

To find a value for p I shall have to differentiate equation (10) to obtain an expression for $\dot{y}$.

$$\dot{y} = \sqrt{\frac{k}{m}}\,p\cos\left(\sqrt{\frac{k}{m}}\,t\right) - \sqrt{\frac{k}{m}}\,q\sin\left(\sqrt{\frac{k}{m}}\,t\right)$$

Putting $\dot{y} = 0$ when $t = 0$ gives

$$0 = \sqrt{\frac{k}{m}}\,p\cos 0 - \sqrt{\frac{k}{m}}\,q\sin 0$$

$$= \sqrt{\frac{k}{m}}\,p - 0$$

So $p = 0$, and in this case the particular solution that models the given starting conditions is

$$y = -h \cos\left(\sqrt{\frac{k}{m}}\,t\right)$$

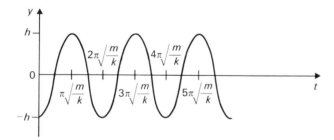

Figure 6

A sketch of this equation is shown in Figure 6. Notice that it is an *inverted* cosine curve—it has its lowest value at $t = 0$. (Check: the scalepan was pushed *down* and then released, so this result is in accord with the initial conditions.)

Let us look at what Figure 6 tells us about the motion. First, notice that it is *repetitive*.

How long does the motion take to repeat?

It repeats after time $2\pi\sqrt{(m/k)}$.

So the model predicts an oscillation which goes on and on, exactly repeating after an interval of time equal to $2\pi\sqrt{(m/k)}$.

The other point that can be discovered from Figure 6 is that the displacement is limited; the scalepan never rises more than h above or falls more than h below its rest position. Moreover, whenever it rises it rises to exactly h above the rest position and whenever it falls it falls to exactly h below the rest position.

My model is, of course, capable of describing other initial conditions—such conditions would lead to different values of the arbitrary constants p and q in the general solution, equation (10). Before I go on to discuss such solutions, I want first to examine in detail the sort of motion my model predicts. I shall do this in the next section.

2.2 Simple harmonic motion

In the last section I arrived at the second-order differential equation

$$\ddot{y} + \frac{k}{m}y = 0 \tag{9}$$

as a model of the motion of a mass on the end of a spring.

Because k and m are both positive, this equation can be written as

$$\ddot{y} + \omega^2 y = 0 \tag{12}$$

where $\omega^2 = k/m$. Because ω^2 must always be positive, equation (12) tells us straight away that the coefficient of y is positive.

Second-order differential equations of this type are very often used as models of motion. They are so common that the motion predicted by differential equation (12) is given a special name. This name is *simple harmonic motion* (SHM for short).

simple harmonic motion

In this section I shall discuss this sort of motion in detail. Remember, however, that I shall be discussing a *model*. SHM is an idealization; it is the predicted motion for an object whose position is described *exactly* by equation (12). When I arrived at differential equation (9) for the oscillating mass I did so on the basis of some modelling suppositions. Later I shall look at the effect of these suppositions on the motion. For now I want to look at the idealized motion—at simple harmonic motion.

What is the general solution of equation (12)? Give it in two forms.

In what follows I shall use the solution in the form $y = r\sin(\omega t + \varepsilon)$. (Remember that it is always possible to manipulate a solution into this form if it has been given in the other form.)

If

$$y = r\sin(\omega t + \varepsilon)$$

then I can use the function of a function rule and differentiate it to obtain $\dot{y}$. Doing this gives

$$\dot{y} = \omega r\cos(\omega t + \varepsilon)$$

Differentiating again gives

$$\ddot{y} = -\omega^2 r\sin(\omega t + \varepsilon)$$

as you saw in SAQ 4.

The general solution is

$$y = p\sin\omega t + q\cos\omega t$$

where p and q are arbitrary constants, or

$$y = r\sin(\omega t + \varepsilon)$$

where r and ε are arbitrary constants.

15

I have sketched graphs of y, $\dot{y}$ and $\ddot{y}$ underneath each other in Figure 7(a), (b) and (c), respectively. From these sketches you can see that

1   when y is at its largest magnitude then $\ddot{y}$ has its largest magnitude in the opposite direction and $\dot{y}=0$;

2   when $y = 0$ then $\ddot{y} = 0$ and $\dot{y}$ has its largest magnitude, alternately one way then the other; and

3   y and $\ddot{y}$ are always in opposite directions.

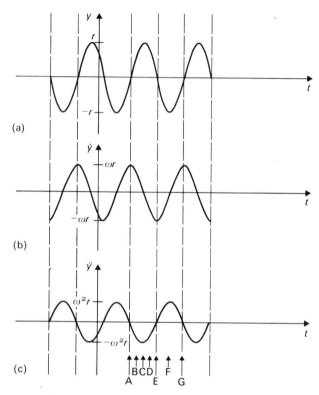

(a)

(b)

(c)

Figure 7

Suppose some object is undergoing the idealized motion I am describing, and that it is oscillating up and down. At the point labelled A on the curves in Figure 7 this object would have zero displacement and zero acceleration ($y = \ddot{y} = 0$) and a positive velocity ($\dot{y} > 0$). It would therefore be moving upwards through its rest position. At point B it is above its rest position ($y > 0$), still moving upwards ($\dot{y} > 0$) but accelerating back down towards its rest position ($\ddot{y} < 0$). Hence its velocity is *decreasing* (negative gradient on $\dot{y}$ curve) At point C it is at its highest point and is at rest ($\dot{y} = 0$). It also has its largest downward acceleration. This causes it to begin to move downwards ($\dot{y} < 0$ at point D). When it passes its rest position ($y = 0$ at point E) it has zero acceleration ($\ddot{y} = 0$), but a downward velocity ($\dot{y} < 0$). Hence, it moves on below its rest position until (point F) it comes to rest ($\dot{y} = 0$). It then has a large upward acceleration ($\ddot{y} > 0$) and so starts to move up again. At point G it begins to repeat its motion.

Notice that there is nothing in this idealized motion to cause a change in the pattern of motion. Simple harmonic motion, once started, continues unchanged forever!

Now that you have discovered something about the nature of simple harmonic motion, let me turn to a discussion of the parameters which characterize it.

In Figure 8 I have sketched on the same axes

$$y = r_1 \sin (\omega t + \varepsilon)$$

$$y = r_2 \sin (\omega t + \varepsilon)$$

$$y = r_3 \sin (\omega t + \varepsilon)$$

where $r_3 > r_2 > r_1$.

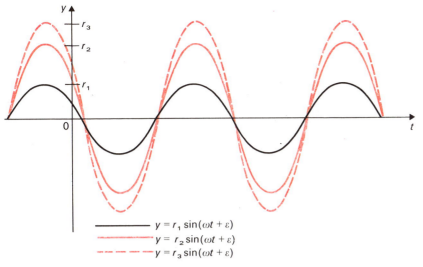

Figure 8

What is the effect of changing the value of r?

The parameter r is referred to as the *amplitude* of the oscillations. The amplitude is the *largest* displacement from the rest position.

amplitude

SAQ 7

SAQ 7

What is the numerical value of the amplitude in each of the following oscillations?

(a) $Y = 3 \sin (2T + \pi/6)$

(b) $Y = 5 \sin (3T + \pi/3)$

(c) $Y = 2.5 \sin (T/2 + 1)$

Notice that changing the amplitude does not affect any feature of the motion other than the maximum displacement. For instance, you can see in Figure 8 that changing the amplitude did not affect the points where the curves crossed the axes.

Now look at Figure 9. Here I have sketched

$$y = r \sin (\omega_1 t + \varepsilon)$$

$$y = r \sin (\omega_2 t + \varepsilon)$$

$$y = r \sin (\omega_3 t + \varepsilon)$$

where $\omega_3 > \omega_2 > \omega_1$.

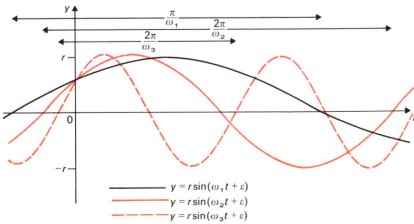

Figure 9

What is the effect of changing the value of ω?

A larger value of ω means that each complete oscillation occurs over a smaller interval of time.

angular frequency

The parameter ω is referred to as the *angular frequency* of the oscillation. Notice that this parameter occurs in the differential equation; the angular frequency is fixed wholly by the equation of motion (that is, the differential equation) and not at all by the way the motion started.

SAQ 8

SAQ 8

What is the numerical value of the angular frequency in each of the following oscillations?

(a) $Y = 3 \sin (2T + \pi/6)$

(b) $Y = 5 \sin (3T + \pi/3)$

(c) $Y = 2.5 \sin (T/2 + 1)$

It was easy to give a physical meaning to the amplitude; it is hard to give one to the angular frequency. It is much easier to give, instead, meaning to two parameters closely related to the angular frequency.

In Figure 9, how long does one complete oscillation take in each of the cases $\omega = \omega_1$, $\omega = \omega_2$ and $\omega = \omega_3$?

For $\omega = \omega_1$, one complete oscillation takes time $2\pi/\omega_1$; for $\omega = \omega_2$ the time is $2\pi/\omega_2$; for $\omega = \omega_3$ the time is $2\pi/\omega_3$.

periodic time, period

The time for one complete oscillation is called the *periodic time* or *period* of the oscillation.

The periodic time is usually given the symbol τ and is related to the angular frequency by

$$\text{periodic time } (\tau) = \frac{2\pi}{\text{angular frequency } (\omega)}$$

The reciprocal of the periodic time (that is, $1/\tau$) is called the *frequency* of the oscillation. The frequency is the number of oscillations completed in a unit of time (usually a second). Thus, if the periodic time is 0.25 seconds, the object will complete 4 oscillations in 1 second and the frequency is $4 \, \text{s}^{-1}$.

frequency

What is the frequency if the periodic time is 2 s?

It is $0.5 \, \text{s}^{-1}$.

Thus, frequency f and periodic time τ are related by $f = 1/\tau$.

I have given the units of frequency as s^{-1}, and certainly as periodic time has dimensions of [time], frequency should have dimensions of $[\text{time}]^{-1}$. By convention, a name is given to the unit of frequency. It is called the *hertz* (Hz). Thus, an oscillation with frequency $10 \, \text{s}^{-1}$ is referred to as having a frequency of 10 Hz.

Because periodic time is related both to angular frequency ω and to frequency f it follows that f and ω are related:

$$\tau = \frac{2\pi}{\omega}$$

$$\tau = \frac{1}{f}$$

so $\omega = 2\pi f$

This shows that ω is just a constant, 2π, times the frequency of the oscillation. Its units are quoted as s^{-1} rather than hertz (Hz). In fact, this is a way of telling an angular frequency from a frequency: angular frequency should have units s^{-1}, frequency Hz. Because some people are rather careless about including the word 'angular' when they are referring to ω, this distinction of units is useful; although, sadly, some people are even careless about the units they use too! But in this course Hz will be used for frequency and s^{-1} for angular frequency.

SAQ 9

SAQ 9

    (a)   An oscillation has a period of 5 s. What are its frequency and its angular frequency?

    (b)   An oscillation has a frequency of 10 Hz. What are its periodic time and its angular frequency?

So a simple harmonic motion given by the differential equation

$$\ddot{y} + \omega^2 y = 0$$

will have angular frequency equal to ω, frequency f equal to $\omega/2\pi$ and periodic time τ equal to $2\pi/\omega$.

There is one more parameter which I need to introduce, and that is phase. In Figure 10 I have sketched

$$y = r\sin(\omega t + \varepsilon_1)$$

$$y = r\sin(\omega t + \varepsilon_2)$$

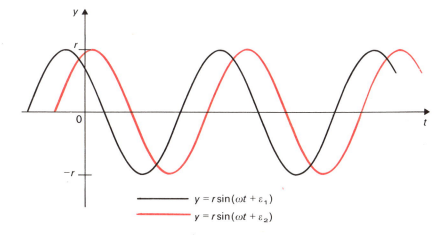

$$y = r\sin(\omega t + \varepsilon_1)$$
$$y = r\sin(\omega t + \varepsilon_2)$$

Figure 10

What is the effect of changing the value of ε?

It shifts the curve along the t axis.

Phase is a measure of the shift along the axis. Clearly phase has to be relative to something fixed. (I have to define a shift relative to something.) I shall therefore define

phase

$$y = r\sin\omega t$$

as having zero phase and

$$y = r\sin(\omega t + \varepsilon)$$

as having phase $+\varepsilon$.* Thus, phase is measured relative to a sine curve which is zero at $t = 0$.

This is not the only definition of phase, but it is the one that will be used in this course.

19

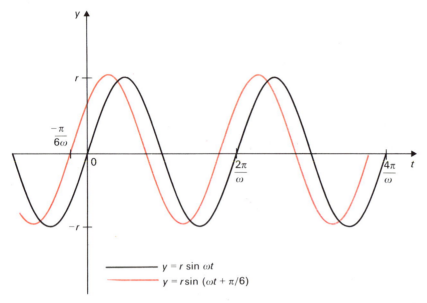

$y = r \sin \omega t$
$y = r \sin (\omega t + \pi/6)$

Figure 11

Figure 11 shows $y = r \sin \omega t$ and $y = r \sin (\omega t + \pi/6)$ on the same axes. You can see that while $y = r \sin \omega t$ cuts the axes at $t = 0$, π/ω, $2\pi/\omega, \ldots, y = r \sin (\omega t + \pi/6)$ cuts the axes at $t = -\pi/(6\omega)$, $5\pi/(6\omega)$, $11\pi/(6\omega), \ldots$ So the presence of the phase $\pi/6$ has shifted the curve along the axis.

SAQ 10

What is the phase in each of the following oscillations?

(a) $Y = 3 \sin (2T + \pi/6)$

(b) $Y = 5 \sin (2T + \pi/3)$

(c) $Y = 2.5 \sin (T/2 + 1)$

You may find it easier to grasp the meaning of phase from the following illustration. Imagine two objects, A and B, each undergoing simple harmonic motion of the same amplitude and frequency. Imagine them side by side and moving up and down.

Now first imagine them both at the highest point of their oscillation together so that they move up and down *together*. The equations describing their motion would have the same phase—whatever value it took, it would be the same for both.

Next imagine that when A is at its highest point B is at its lowest. As A moves down, B will move up; as B moves down A will move up. The two equations describing their motion would *not* have the same phase; in fact, the two phases would differ by π, although that detail need not concern you unduly.

Finally, imagine that B is following A so that B reaches its highest point just fractionally after A, and so on throughout the motion. The two equations describing their motion would have nearly-equal phases; but the phases would not be *exactly* equal because A and B are not *quite* moving together.

So while frequency (or periodic time) tells you about how rapid the oscillations are and amplitude tells you how large they are, phase enables you to pinpoint where the oscillating object is at various times, or to compare the relative positions of two oscillating objects at various times.

One final point—remember that the frequency is determined by the differential equation itself, but that the amplitude and phase are determined by the way the motion started.

20

Write down the amplitude, phase, angular frequency, frequency and periodic time of each of the following sinusoids. (Displacements are in millimetres, time is in seconds.)

(a) $Y = 3 \sin 0.2T$

(b) $Y = 5 \sin 2T + 12 \cos 2T$ (Hint: use equations (5) and (6) to rewrite this in the appropriate format first.)

To summarize, then, simple harmonic motion is an idealized motion with the following properties:

1   It is described by the differential equation

$$\ddot{y} + \omega^2 y = 0$$

2   It is repetitive oscillation with period $2\pi/\omega$, frequency $\omega/2\pi$ and angular frequency ω.

3   The solution of the differential equation is of the form

$$y = r \sin(\omega t + \varepsilon)$$

where r is the amplitude and ε the phase, these two parameters being constant and determined by information about how the motion started.

2.3   Simple harmonic motion and motion in a circle

You may have noticed an analogy with some of the features of circular motion introduced in Unit 9. There ω was called the angular velocity and τ the time period to complete one revolution of the circle. Here ω is angular frequency and τ is periodic time. Is there some relationship?

The relationship is between the co-ordinates of the position of the object going round in a circle and simple harmonic motion. You may remember that for a car going in a circle of radius r at angular velocity ω, the co-ordinates are

$$x = r \cos(\omega t + \varepsilon)$$

$$y = r \sin(\omega t + \varepsilon)$$

where the car started at an angle ε with the x-axis.

If you were to sketch the way the y co-ordinate varies with t you would find it identical to Figure 7(a), the sketch of how displacement varies with time in SHM. The sketch for the x co-ordinate would be identical except for a shift

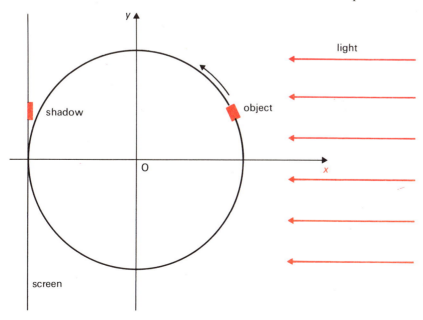

Figure 12

21

along the t-axis; that is, identical except for a phase shift. Thus, the position co-ordinates of the motion both follow simple harmonic motion. One way to think of this is to note that if you built a model car to run on a circular track and then beamed light from one side onto it with a screen parallel to the y-axis (Figure 12) the shadow would move across the screen in an exactly analogous manner to the way an object follows simple harmonic motion (for example the scalepan of the spring balance). So although an object moving in a circle does not follow simple harmonic motion, each of its two position coordinates does follow simple harmonic motion. This explains the similar features in the two types of motion.

2.4 Particular solutions of the differential equation in the model

Now that you have learned something about simple harmonic motion and the parameters which characterize it, you are in a position to apply this knowledge to the model of the motion of a spring balance which I introduced in Section 2.1.

The differential equation modelling the motion is

$$\ddot{y} + \frac{k}{m}y = 0 \tag{9}$$

with general solution

$$y = r \sin\left(\sqrt{\frac{k}{m}}t + \varepsilon\right) \tag{11}$$

where r and ε are arbitrary constants.

What is the angular frequency of this motion?

It is $\sqrt{(k/m)}$.

What is the periodic time?

It is $2\pi\sqrt{(m/k)}$. (Note: $1/\sqrt{(k/m)} = \sqrt{(m/k)}$.)

Notice that the periodic time depends *only* on the spring constant k and the mass m. The model therefore predicts that the oscillations of a given mass on a given spring will always have the same periodic time. Further, the periodic time can always be predicted from a knowledge of k and m.

Let me suppose that, for a particular spring balance, placing a mass of $100\,\text{g} = 0.1\,\text{kg}$ into the scalepan extends the spring by $5\,\text{cm} = 0.05\,\text{m}$. The spring constant was defined by equation (7)

$$f = -ky$$

where f is the force acting when there is a displacement y. Re-arranging this equation gives

$$k = -\frac{f}{y}.$$

Here, $f = 0.1 \times 9.8\,\text{N}$ and $y = -0.05\,\text{m}$. (The minus sign arises because the displacement is downwards.) So

$$k = -\frac{0.1 \times 9.8}{-0.05}\,\text{N m}^{-1}$$

$$= 19.6\,\text{N m}^{-1}$$

Note the dimensions; k is in units of newtons per metre.

The periodic time is

$$\tau = 2\pi \sqrt{\frac{m}{k}}$$

$$= 2\pi \sqrt{\frac{0.1}{19.6}} \, \text{s}$$

$$= 0.45 \, \text{s} \qquad \text{(to two significant figures)}$$

SAQ 12

SAQ 12

The contents of the scalepan are changed so that the total mass of the scalepan and contents is 400 g. Find the periodic time of the oscillations.

Although it is possible to predict the periodic time (and therefore, of course, the frequency and angular frequency) of the oscillations from either the differential equation or its general solution, it is necessary to have a particular solution to find values for the amplitude and phase. This is because the amplitude and phase depend on the way the motion was started.

You saw one example of a particular solution in Section 2.1. There the scalepan was pulled down a distance h and then gently released. This led to the particular solution

$$y = -h \cos \left(\sqrt{\frac{k}{m}} t \right)$$

What are the amplitude and phase of this oscillation?

The amplitude is h (amplitudes are always positive since they are greatest displacements). The phase cannot be found by inspection. I can, however, use equation (6),

$$\tan \varepsilon = \frac{p}{q}$$

Here, $q = -h$ and $p = 0$ (there is no sine term in the solution). Hence

$$\tan \varepsilon = -\frac{h}{0}$$

$$= -\infty$$

Now in the range 0 to 2π, $\tan \varepsilon$ is infinitely large and negative when $\varepsilon = \pi/2$ or $3\pi/2$. (Check this using the graph of $\tan \theta$ in your *Handbook*, if necessary.) I shall have to decide which value of ε to take—in other words, which is identical to $y = -h \cos \{\sqrt{(k/m)}t\}$: $y = h \sin \{\sqrt{(k/m)}t + \pi/2\}$ or $y = h \sin \{\sqrt{(k/m)}t + 3\pi/2\}$? I know that y should be $-h$ at $t = 0$. Only $y = h \sin \{\sqrt{(k/m)}t + 3\pi/2\}$ fulfils that condition, so $\varepsilon = 3\pi/2$ and this is the value of the phase in this case.

In this particular case the scalepan had a displacement but no velocity when $t = 0$. Frequently, of course, the scalepan will be given both an initial displacement and an initial velocity when its oscillations start.

Example

A model of a particular spring balance with particular scalepan contents leads to the differential equation

$$\ddot{Y} + 4Y = 0$$

where Y metres is the displacement at time T seconds. Find how Y varies with T, given that the oscillations started when the scalepan was pushed down by $2\,\text{cm} = 0.02\,\text{m}$ from its rest position and given an initial velocity downwards of $3\,\text{cm}\,\text{s}^{-1} = 0.03\,\text{m}\,\text{s}^{-1}$.

The general solution of this differential equation is, in the form which is easier to handle when a particular solution is to be found.

$$Y = P \sin 2T + Q \cos 2T$$

where P and Q are arbitrary constants. The initial conditions can be expressed in mathematical form as $Y = -0.02$ and $\dot{Y} = -0.03$ at $T = 0$.

Why are both Y and $\dot{Y}$ negative at $T = 0$?

Because both the displacement and the initial velocity are *downwards*.

Using $Y = -0.02$ at $T = 0$ gives

$$-0.02 = P \sin 0 + Q \cos 0$$
$$= 0 + Q$$

hence

$$Q = -0.02$$

Differentiating the general solution gives

$$\dot{Y} = 2P \cos 2T - 2Q \sin 2T$$

and putting in $\dot{Y} = -0.03$ at $T = 0$ gives

$$-0.03 = 2P \cos 0 - 2Q \sin 0$$
$$= 2P - 0$$

Hence $P = -0.015$ and the particular solution is

$$Y = -0.015 \sin 2T - 0.02 \cos 2T$$

What is the periodic time of this oscillation?

It is $2\pi/2\,\text{s} = \pi\,\text{s} = 3.14\,\text{s}$ (to three significant figures).

What is the amplitude of this oscillation?

The amplitude can be found from equation (5). It is

$$\sqrt{\{(-0.015)^2 + (-0.02)^2\}}\,\text{m}$$
$$= \sqrt{(0.000225 + 0.0004)}\,\text{m}$$
$$= 0.025\,\text{m}$$

SAQ 13

Find the particular solution of

$$\ddot{Y} + 9Y = 0$$

given $Y = 0.01$, $\dot{Y} = 0.03$ at $T = 0$.

What are the amplitude and the periodic time for this motion, given Y is a displacement in metres and T time in seconds?

SAQ 14

A scalepan and its contents have mass 0.5 kg. The spring constant of the spring is 18 N m^{-1}.

(a) Write down the differential equation for these oscillations and find its general solution.

(b) Find the particular solution if the oscillations begin 0.05 m below the scalepan's rest position with an upwards velocity of 0.03 m s^{-1}.

(c) Find the amplitude and periodic time of these oscillations.

2.5   How does the model compare with reality?

In the preceding sections I have found a model for the oscillations of the scalepan of a spring balance. I found that my model predicted a particular type of motion called simple harmonic motion. I want now to examine how well simple harmonic motion tallies with the motion I expect of a real spring balance.

In one respect the model is certainly wrong. It predicts that, once the scalepan is set in motion, it will continue to oscillate forever with a fixed frequency and amplitude.

This is not in accord with reality. What actually happens?

Real oscillations die away; their amplitudes are *not* constant but become progressively smaller. I shall describe such motion as a *decaying* or *damped oscillation*.

decaying oscillation
damped oscillation

To see why my model does not predict the expected decay of real oscillations, I shall return to the suppositions on which it was based. Two suppositions seem trivial. They were that the motion was vertical and that the support was rigid. It is hard to see in them any cause for the lack of decay of oscillations.

A third supposition was that any resistance to the motion was negligible. Now here I do have a supposition that may well make the difference between decaying and non-decaying oscillations. You may remember from Unit 11 that, while neglecting air resistance for a falling body led to a model where the object goes on accelerating until it reaches the ground, taking air resistance into account predicts a terminal velocity for a falling object. So resistive forces can have quite a significant effect.

Let us look at the sort of effect which resistance might have. It provides a force which acts in the opposite direction to the one in which the object is moving.

Would you expect this to lead to a decaying oscillation?

It does seem likely, at least. To be certain I need to make a new model which incorporates forces of this sort. To simplify my model, I shall suppose that I may lump together all sources of resistance and model them by a single force whose magnitude is proportional to velocity but which acts in the opposite direction. I shall denote the constant of proportionality by r so that the resistive force $= -r\dot{y}$.

The total force on the scalepan is therefore $-r\dot{y} - ky$ (remember that $-ky$ is the force we had before, in the model which neglected resistance), and this produces an acceleration $\ddot{y}$. Using Newton's second law gives

$$m\ddot{y} = -r\dot{y} - ky$$

which can be rewritten

$$m\ddot{y} + r\dot{y} + ky = 0 \qquad (13)$$

This is a second-order differential equation of a type you met before. In fact, you learned in Unit 14 that under certain circumstances such a differential equation can have a solution which is a decaying oscillation.

What are the conditions for the differential equation

$$a\frac{d^2y}{dx^2} + b\frac{dy}{dx} + cy = 0$$

to have a solution which is a decaying oscillation?

They are:

(i) $b^2 < 4ac$

(ii) $b > 0$

In equation (13) we have m, r and k instead of a, b and c. You already know that k and m are positive constants, and I defined r in such a way that it, too, is a positive constant. Thus, $r > 0$. Provided r^2 is small enough that $r^2 < 4mk$, the solution of equation (13) is, from Unit 14 or your *Handbook*,

$$y = \exp\left(\frac{-rt}{2m}\right)\left[p\sin\left\{\frac{\sqrt{(4mk - r^2)}}{2m}t\right\} + q\cos\left\{\frac{\sqrt{(4mk - r^2)}}{2m}t\right\}\right] \qquad (14)$$

where p and q are arbitrary constants. Figure 13 shows the general shape of the graph of such a solution. The greatest amplitude and the phase of the motion will depend on the initial conditions, but whatever the initial conditions the solution is always a sinusoid whose successive peaks fall on a decaying exponential curve.

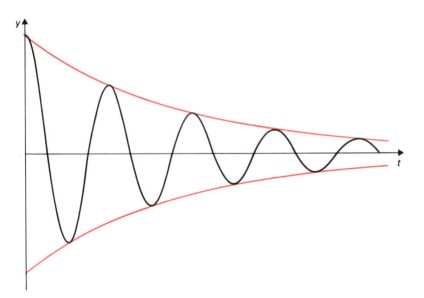

Figure 13

Is the angular frequency of the oscillations still constant in such a model?

You can find the angular frequency of the oscillations by looking at the coefficient of t in the sine and cosine terms. This coefficient is $\sqrt{(4mk - r^2)}/2m$. Now m, k and r are all constants, so the angular frequency is also constant. Hence, although the amplitude decays the oscillations still have a constant periodic time and frequency.

Does the angular frequency predicted for the damped oscillation equal that which would be predicted in a model which ignores resistive forces?

The expression for the angular frequency of damped oscillations

$$\frac{\sqrt{(4mk - r^2)}}{2m}$$

can be written thus

$$\sqrt{\left(\frac{4mk - r^2}{4m^2}\right)}$$

$$= \sqrt{\left(\frac{k}{m} - \frac{r^2}{4m^2}\right)}$$

The expression for the angular frequency of undamped oscillations is

$$\sqrt{\frac{k}{m}}$$

so you can see that the angular frequency predicted for the damped oscillations is *smaller* than the angular frequency predicted for undamped oscillations by ignoring resistive forces. This means, of course, that the periodic time is *larger*. How much smaller the angular frequency is depends on the size of r; the larger the damping force the larger r and the greater the reduction in angular frequency.

What is the angular frequency of a motion described by the differential equation

$$\ddot{Y} + 2\dot{Y} + 2Y = 0$$

where Y is displacement in centimetres at time T seconds?

Here $K = 2$, $M = 1$ and $R = 2$. Hence the angular frequency is

$$\frac{\sqrt{(4 \times 1 \times 2 - 2)}}{2 \times 1}\,\text{s}^{-1}$$

$$= \frac{\sqrt{4}}{2}\,\text{s}^{-1}$$

$$= 1\,\text{s}^{-1}$$

What angular frequency would be predicted if resistive forces were ignored?

The motion would then be modelled by the differential equation

$$\ddot{Y} + 2Y = 0$$

with angular frequency $\sqrt{2}\ \text{s}^{-1} = 1.41\ \text{s}^{-1}$ (to three significant figures). This is, as expected, larger.

The fact that oscillations die away is a desirable property in a spring balance. Were a spring balance to oscillate indefinitely whenever it was disturbed, it would not be a very useful device! In such a device resistance to motion is sometimes built in so that measurements may be taken quickly.

In Section 4 I shall look at another mass–spring system. You will see that there resistive forces are also desirable and are provided. First, however, in Section 3 I shall look at another model which predicts oscillations.

2.6 Summary

When resistive forces are neglected, the motion of a mass on the end of a spring may be modelled by the second-order differential equation

$$m\ddot{y} + ky = 0$$

Using this model, the position co-ordinate is given by

$$y = p\sin\left(\sqrt{\frac{k}{m}}t\right) + q\cos\left(\sqrt{\frac{k}{m}}t\right)$$

where p and q are arbitrary constants, or by

$$y = r\sin\left(\sqrt{\frac{k}{m}}t + \varepsilon\right)$$

where r and ε are arbitrary constants.

The predicted motion is periodic with periodic time τ equal to $2\pi\sqrt{(m/k)}$, frequency f equal to $(1/2\pi)\sqrt{(k/m)}$ and angular frequency ω equal to $\sqrt{(k/m)}$.

The amplitude of the predicted motion is given by r and the phase by ε.

The motion predicted by the model is an example of simple harmonic motion, which is generally described by the differential equation

$$\ddot{y} + \omega^2 y = 0$$

where ω is the angular frequency of the motion. Simple harmonic motion is always predicted in a situation where the acceleration on an object is modelled as being proportional to its displacement but in the opposite direction.

This simple model of a mass on a spring does not accord with reality in that it predicts oscillations of a fixed amplitude which go on indefinitely. In practice, oscillations which die away are to be expected. Incorporation into the model of a resistance force leads to a prediction of decaying oscillations.

3 MODELLING THE OSCILLATIONS OF A PENDULUM

You can easily make a simple pendulum. All you need is a small weight—a coin, a nut or a key would do perfectly—and a piece of thread, perhaps a metre or so long: a builder's plumb line would do admirably. You fix the weight to one end of the thread and tie the other to a firm support so that the weight hangs free. If you then pull the weight a little way to one side and let it go, it will oscillate from side to side below the support in a motion that is quite closely simple harmonic. The motion will continue for some time, but it will gradually die away leaving the weight at rest directly below the support.

To construct an equation which will describe the motion of the weight, which I shall now call the 'bob' of the pendulum, I first need a mathematical model.

I shall suppose that the thread is of fixed length, l, and is firmly secured, Figure 14. If the bob is small, I can consider it to be 'a point mass', so that I need not consider its shape. I shall neglect the mass of the thread, taking the mass, m, of the bob as the only one to enter the calculations.

With these restrictions, my model is that of a point mass m supported by a light inextensible thread of length l moving in the plane represented in Figure 14 by the plane of the page. In this model, the position P of the bob may be represented by the angle θ, the angle the thread makes with the vertical. The problem I want to pose is that of finding an equation which will represent the variation of θ with time, t.

Consider the forces acting on the mass m. They are the weight, mg, acting downwards and the tension f in the string acting along the length of the string. In the position θ the weight of the bob has one component $mg \cos \theta$ acting along the line of the thread which tends to extend it. Another component acts perpendicular to the line of the string. It has the value $mg \sin \theta$ and, since it tends to bring the bob towards the central position, it is called the *restoring force*.

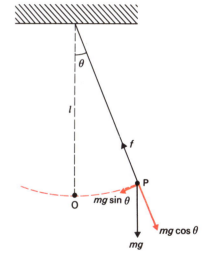

Figure 14

Since the string is of fixed length, the arc through which the point P moves is circular. I can describe its velocity, therefore, in terms of the angle θ. Using the dot notation, I can write the angular velocity of P as $\dot{\theta}$, its speed is then given by $l\dot{\theta}$ (from Unit 9, $v = r\omega$). This is regarded as positive for *increasing* θ.

I can therefore write the acceleration of P along its circular path as the rate of change in the product $l\dot{\theta}$. As the length l is constant, this rate is given by

$$l\frac{\mathrm{d}(\dot{\theta})}{\mathrm{d}t} = l\ddot{\theta}$$

The component of force $mg \sin \theta$ acts at right-angles to the supporting thread and so it is also along the circular path of the bob in the direction of *decreasing* θ. It therefore has the opposite sign to the acceleration. Using Newton's second law

$$-mg \sin \theta = m(l\ddot{\theta})$$

This equation relates the position, θ, of the bob to its acceleration, $\ddot{\theta}$. Notice that the quantities related by the equation are *angles*.

The mass cancels from the equation to give

$$\ddot{\theta} + \frac{g}{l}\sin\theta = 0 \qquad (15)$$

What kind of equation is this?

This is a second-order differential equation relating θ and t.

Is it an example of the simple harmonic equation?

No.

My theoretical model has not led me to the simple harmonic equation, but rather to differential equation (15). Unfortunately this differential equation is not readily soluble—there is no convenient, well-known function which satisfies this differential equation, and therefore computational techniques are needed to solve it. However, there is a simplification I can make provided I restrict the range of my model slightly.

Notice that differential equation (15)

$$\ddot{\theta} + \frac{g}{l}\sin\theta = 0 \qquad (15)$$

is very like the equation

$$\ddot{\theta} + \frac{g}{l}\theta = 0 \qquad (16)$$

which defines simple harmonic variation of θ with t. So can I relate θ to $\sin\theta$?

Table 1   Values of sin θ for small angles

θ	1°	2°	3°	4°	5°	10°	15°
$\sin\theta$	0.01745	0.03490	0.05234	0.06976	0.08716	0.1736	0.2588
θ/rad	0.01745	0.03491	0.05236	0.06981	0.08727	0.1745	0.2618

Look at Table 1. There, $\sin\theta$ and θ (in radians) are given for a range of values of θ. You can see that $\sin\theta \approx \theta$, the approximation being very close indeed for small angles and accurate to 1 part in 200 or so even for 10°. Figure 15 makes the same point—even at 30°, θ and $\sin\theta$ differ only by about 4 per cent.

Therefore provided I restrict my model to the case where the angle of swing of the pendulum is no more than about 0.2 radians ($\sim 10°$) each side of the rest position I can use differential equation (16)

$$\ddot{\theta} + \frac{g}{l}\theta = 0$$

to model the motion—and this is an equation whose solution can readily be found. It is

$$\theta = r\sin\left(\sqrt{\frac{g}{l}}\,t + \varepsilon\right)$$

Why might a simple pendulum be used in the mechanism of a clock?

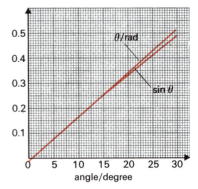

Figure 15

Because the model pedicts a fixed periodic time. In fact, adjusting l appropriately can give an expected period of exactly 1 second, because the model predicts a period of $2\pi\sqrt{(l/g)}$.

SAQ 15

About how long must a simple pendulum be if its oscillations are to have a period of one second? (Use $g = 9.8\,\mathrm{m\,s^{-2}}$ and give your answer to two significant figures.)

Do the suppositions in the model suggest any precautions a clockmaker may need to take?

SAQ 16

The differential equation for the motion of the pendulum bob given above has a solution which relates the angle θ to time t. If the *horizontal* displacement of the bob from its rest position is x will the variation of x with time be simple harmonic?

SAQ 17

A simple pendulum is of length 1 m. The bob is displaced 15 cm horizontally from its rest position below the support and gently released. Find an equation for the subsequent motion using the approximation introduced in this section.

3.1 How does the model compare with reality?

In any real pendulum, the oscillations tend to die away. My model predicts oscillations which go on indefinitely because I have ignored resistance to motion. I can include resistive forces in my model in much the same way as I did for the spring balance model in Section 2.5.

I shall again suppose the resistance to be proportional to velocity, but in the opposite direction. For a pendulum, the angular velocity $\dot{\theta}$ is a constant multiple of velocity and thus I shall have a resistive force of $-r\dot{\theta}$, where r is the proportionality constant.

Adapting the argument of the preceding section to incorporate this additional force gives

$$\text{force} = mg \sin\theta - r\dot{\theta}$$

but force $=$ mass $\times$ acceleration

$$= ml\ddot{\theta}$$

Hence $ml\ddot{\theta} = -mg \sin\theta - r\dot{\theta}$

I shall again confine my attention to small amplitudes so that I may use the approximation $\sin\theta = \theta$. The differential equation thus becomes

$$ml\ddot{\theta} + r\dot{\theta} + mg\theta = 0 \qquad (17)$$

Solutions of this differential equation are found in just the same way as for differential equation (13) for a mass on a spring, as the next two SAQs show

SAQ 18

(a) Under what conditions would you expect a solution of equation (17) to be a decaying oscillation?
(b) Write down the general solution in that case.

SAQ 19 (difficult)

A pendulum of length 1 m and with a bob of mass 100 g swings in a medium which offers a resistive force of 0.1 N when the angular velocity is $0.5\,\mathrm{rad\,s^{-1}}$.

(a) Write down the differential equation which relates the angular displacement θ to time T seconds. (Take $g = 10\,\mathrm{m\,s^{-2}}$).
(b) Find the general solution of this differential equation.

(c) Find the particular solution if the bob is hit when in its rest position such that it is given an angular velocity of $0.3\,\text{rad}\,\text{s}^{-1}$.

(d) Sketch this solution, showing any important features of the motion.

While the $\sin\theta = \theta$ approximation is an excellent one for small angles, the motion thus predicted for the point P of the model differs sufficiently from the actual motion of a pendulum that, for one application at least, this solution of the differential equation was inadequate and a higher level of approximation was sought.

The problem is now of historical interest. Clockmakers wanted a reasonable amplitude of swing on their pendulums, because this makes it easier to set the escapement. There was a consequent interest in using amplitudes which were not really 'small'.

That the movement of the pendulum is not simple harmonic at large amplitudes does not really matter to a clockmaker. He is, however, concerned if the periodic time of the pendulum changes with amplitude; if, for example, his clocks, when freshly wound, have larger periods than when wound down, for then the timekeeping would vary.

Now the model I have just derived, which takes resistance to motion into account, predicts decaying oscillations and so the amplitude is not fixed but decreases with time. It also predicts a fixed period of oscillation—but this prediction only holds if $\sin\theta = \theta$.

Figure 15 showed how $\sin\theta$ and θ compare for $0 \leqslant \theta \leqslant 30°$. At the larger angles they differ by about 4 per cent, which is significant by clockmaking standards—it is more than two minutes per hour. Now the restoring force on a pendulum varies with angle, following the graph of $\sin\theta$, while the graph of θ shows how the force *should* vary if the motion is to be sinusoidal. Hence, at large amplitudes the forces on a pendulum are too small to make equation (17) an accurate model. Because the forces are too small the accelerations are too small, and the time taken for a complete oscillation is greater than for a small-angle oscillation.

What does this mean?

It means that as the amplitude of the swing decreases with time-since-winding, so the periodic time will decrease and the clock will go faster. This is a case where the damping force produced by resistance to the motion is a nuisance.

It is possible to refine the mathematical treatment of the pendulum problem to the point at which the model will give a precise estimate for the variation of periodic time with amplitude; but who wants a clock with a correction table telling you how fast or slow it should be so many days on from its last winding?

Clockmakers, by the way, never did try to make corrections to the equations to allow for variations in amplitude of the swing pendulum. Their interest was the practical one of obtaining a fixed period for the motion of a pendulum.

They therefore hung on to the differential equation for true simple harmonic motion and set about creating a pendulum suspension for which this equation is an accurate model. Their method was simplicity itself: the restoring force was increased by mounting the pendulum between curved jaws (Figure 16), effectively shortening the pendulum for that part of its motion which is at a large displacement. The curve of the jaws was of a particular shape, a cycloid, but the construction was simple and the method effective. Modern pendulum clock escapements make the device largely irrelevant by closely controlling the amplitude of swing.

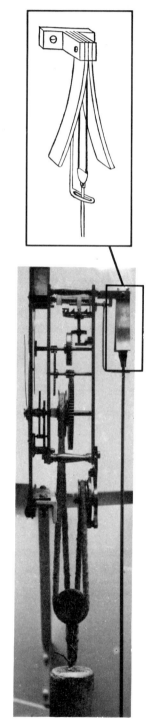

Figure 16 In Huygens' clock the upper part of the pendulum's suspension is flexible and ranges between curved jaws (inset). (Courtesy Science Museum, London.)

4 MODELLING THE OSCILLATIONS OF AN ACCELEROMETER

The effect of resistive forces has already been examined to some extent in Sections 2.5 and 3.1. There you have seen that such forces produce oscillations of decreasing amplitude and of a fixed frequency. This frequency is less than the frequency of oscillation if the resistive forces could be removed. Let me look again at equation (13), the differential equation I derived for the motion of the scalepan of a spring balance when resistance is taken into account

$$m\ddot{y} + r\dot{y} + ky = 0 \tag{13}$$

I have talked about the solution of this differential equation as being a decaying oscillation, but, of course, this form of solution only occurs if r is small enough for $r^2 < 4mk$ to be true. In a spring balance the resistive forces are usually small and r^2 is indeed likely to be less than $4mk$. But sometimes oscillations, even decaying ones, are unwelcome and a damping mechanism is deliberately introduced into a physical system in order to increase r, so that the oscillations die away very quickly or even do not occur at all.

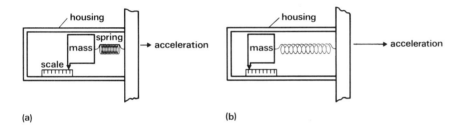

(a) (b)

Figure 17(a) Small acceleration:
(b) large acceleration

One example of this deliberate damping is in an instrument which measures acceleration. It is called a mass–spring accelerometer. In its simplest form it consists of a mass at the end of a horizontal spring. In the case of a vertical spring it is the vertical acceleration due to gravity acting on the mass hung on it which causes the spring to extend. So a *horizontal* spring will be extended by a *horizontal* acceleration acting on the mass on the end of it, and the larger the acceleration the greater the extension. Thus, when the device to which the accelerometer is attached is accelerated the spring extends. This is illustrated in Figure 17. The scale can be calibrated to show the size of the acceleration.

There is a problem when the acceleration suddenly changes in size. Can you see what this is?

If the acceleration begins suddenly the spring will be extended suddenly and this causes the mass to oscillate. It is then very difficult to take a reading. This problem is overcome by damping the motion, as shown in Figure 18. The piston in the oil-filled cylinder provides a large force which always resists the motion of the mass.

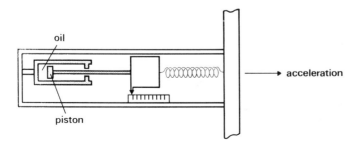

Figure 18

Differential equation (13) is still an appropriate model; even though the spring is now horizontal instead of vertical:

$$m\ddot{y} + r\dot{y} + ky = 0 \qquad (13)$$

In this case r represents the combined resistance to motion of the air (a small force), of the spring (a smallish force) and of the piston in the oil (quite a large force). Hence here r is not necessarily such that $r^2 < 4mk$. In fact, a value of r can be selected by choosing an oil of appropriate 'stickiness'. As you might expect a thin oil will provide much less resistance to motion than a thick, sticky one.

What would happen to the behaviour of the mass when suddenly accelerated if different oils were used one by one, each one increasing the value of r?

First, consider a very 'light' oil, such that r^2 is less than $4mk$. Here equation (14) from Section 2.5 gives the solution:

$$y = \exp\left(\frac{-rt}{2m}\right)\left[p\sin\left\{\frac{\sqrt{(4mk - r^2)}}{2m}t\right\} + q\cos\left\{\frac{\sqrt{(4mk - r^2)}}{2m}t\right\}\right] \quad (14)$$

where p and q are arbitrary constants. With heavier oils, r is larger. As r increases there are two effects which can be predicted from equation (14). One is that the exponential decay is faster, because the time constant of the decay is $2m/r$ and this decreases as r increases. The other is that the angular frequency of oscillation becomes smaller because if r increases then $\dfrac{\sqrt{(4mk - r^2)}}{2m}$ decreases. This means, of course, that the periodic time increases. Figure 19 shows the nature of the oscillations about the final, rest, position in these two cases. You can see that as r gets larger the oscillations are quickly damped out.

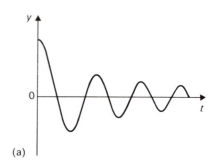

(a)

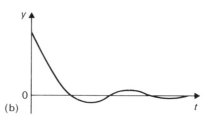

(b)

Figure 19   r has a larger value in (b) than in (a)

This situation of decaying oscillations is sometimes described by the term *damped simple harmonic motion*. Its characteristics are that the frequency is fixed as in true simple harmonic motion, but that the amplitude decays exponentially.

damped simple harmonic motion

If the damping is increased still further then eventually r^2 will cease to be less than $4mk$. It will become first equal to it, and then greater than it. You examined the solution of a differential equation subject to similar conditions in Section 3.3 of Unit 14. There, the differential equation was

$$a\frac{d^2y}{dx^2} + b\frac{dy}{dx} + cy = 0$$

and the conditions were $b^2 = 4ac$ and $b^2 > 4ac$.

What were the solutions under these two conditions?

When $b^2 = 4ac$ the solution is

$$y = \exp\left(\frac{-bx}{2a}\right)(px + q)$$

and when $b^2 > 4ac$ the solution is

$$y = p\exp\left\{\frac{-b + \sqrt{(b^2 - 4ac)}}{2a}x\right\} + q\exp\left\{\frac{-b - \sqrt{(b^2 - 4ac)}}{2a}x\right\}$$

In both cases, p and q are arbitrary constants.

In the case of the accelerometer the solutions are exactly analogous, with t instead of x, m instead of a, r instead of b and k instead of c. Figure 20 shows what would happen in the rather special case where $r^2 = 4mk$. The mass moves to its final position without overshooting it.

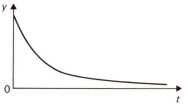

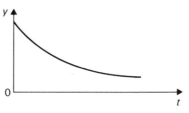

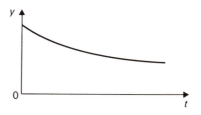

Figure 20

Figure 21

Of course, setting r^2 precisely equal to $4mk$ is not possible. Figures 21(a) and (b) show what happens when $r^2 > 4mk$. In Figure 21(a), r^2 is only just greater than $4mk$. In Figure 21(b), r^2 is very much greater than $4mk$. You can see the effect of making r larger; the mass takes longer to reach its rest position. Once again, there is no overshoot of this rest position.

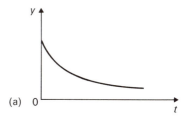

(a)

Sometimes the motion where oscillations take place (such as in Figure 19) is referred to as *underdamped*, motion like that in Figure 20 is referred to as *critically damped* and motion like that in Figure 21 as *overdamped*. These terms can be confusing, however, because a degree of damping which is sufficient for some purposes may be insufficient for others. This course does not use them.

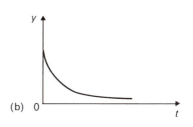

(b)

If you were a manufacturer of a mass–spring accelerometer, which type of behaviour would you want of the five illustrated in Figures 19, 20 and 21? Take it that the timescales along the axes are identical in all five cases.

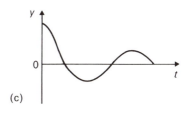

(c)

The obvious answer is the behaviour illustrated in Figure 20. This motion corresponds to the criterion of the most rapid return to the rest position without any overshoot. Suppose, however, that your criterion was simply a rapid approach to the rest position. Is the behaviour of Figure 20 still the best to choose?

As it happens, the answer is no. It can be shown that the motion illustrated in Figure 19(b) gives a more rapid response, because if $\sqrt{(r^2/4mk)} \approx 0.7$ the mass gets close to, and stays close to, its final position more rapidly. Figures 19(b) and 20 show this, if you look closely.

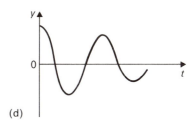

(d)

So, if the instrument was constructed so that any overshoot of the final rest position caused damage, it would be necessary to ensure that r^2 was at least equal to $4mk$. If, however, overshoot was permissible and a rapid approach to very close to the final position was desirable then it would be better to ensure that $\sqrt{(r^2/4mk)} \lesssim 0.7$.

SAQ 20

SAQ 20

Figure 22 shows various examples of the behaviour of a mass-spring accelerometer with different damping. The timescales are the same. Place Figure 22 parts (a) to (e) in order of increasing value of r and state in each case whether r^2 is greater than, less than or equal to $4mk$.

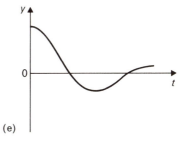

(e)

Figure 22

Television programme 9, *Braking for Safety*, examines this type of accelerometer as well as some other types which are used in car braking tests.

35

5 CONCLUSION

In Section 3 of Unit 14 I introduced a family of second-order differential equations of the type

$$a\frac{d^2y}{dx^2} + b\frac{dy}{dx} + cy = 0 \qquad (a \neq 0)$$

and showed you how to find the general solution of any member of this family. Since then I have been concerned to show you how such differential equations arise in modelling situations. In the space available I have only been able to discuss a few of the many situations where they do occur; for instance I have not touched on the subject of electric circuits, where models describing the voltages in circuits containing various combinations of certain types of electrical components can lead to members of this family of differential equations. What I have tried to do is to show you enough about how they arise, how they are solved and what these solutions mean in practice to help you tackle models involving these differential equations if you meet them in your future studies.

In particular I have tried to show you that although a simple model may lead to a prediction of oscillations of fixed amplitude which continue for ever (simple harmonic motion), in practice there is always some degree of damping which will cause these oscillations to die out. If this damping is large enough oscillations may not occur.

When the above differential equation is used to model the motion of a damped system it is possible to predict readily whether the motion will be oscillatory or will take the form of a gradual return to the rest position. To make this prediction, it is necessary only to examine the relative sizes of the constant parameters in the differential equation.

So these two units have tried to give you

1 a knowledge of the solutions of the family of differential equations

$$a\frac{d^2y}{dx^2} + b\frac{dy}{dx} + cy = 0 \qquad (a \neq 0)$$

   and an ability to find the solution of any member of the family.

2 an understanding of how to apply this knowledge in a modelling situation.

With these two skills you have useful tools for investigating many technological problems.

SUMMARY OF THE UNIT

The general solution of the differential equation

Section 1.1

$$\ddot{y} + \omega^2 y = 0$$

may be written either in the form

$$y = p \sin \omega t + q \cos \omega t$$

or in the form

$$y = r \sin (\omega t + \varepsilon)$$

where

$$p = r \cos \varepsilon \qquad r = \sqrt{(p^2 + q^2)}$$

$$q = r \sin \varepsilon \qquad \tan \varepsilon = q/p$$

The motion of a *mass m hanging on the end of a spring* may be modelled by the differential equation

Section 2.1

$$\ddot{y} + \frac{k}{m}y = 0$$

where k is a property of the spring called the *spring constant*.

Simple harmonic motion is the name given to the idealized motion which is described precisely by the differential equation

Section 2.2

$$\ddot{y} + \omega^2 y = 0$$

The solution of this differential equation is

$$y = r \sin (\omega t + \varepsilon)$$

r is called the *amplitude* of the motion, ω is called the *angular frequency* and ε the *phase*. Two other parameters related to ω are the frequency ($f = \omega/2\pi$) and the periodic time ($\tau = 2\pi/\omega$). The angular frequency is determined by the differential equation, but the amplitude and phase are determined by the way the motion started (that is, by the initial conditions).

When an object moves in a *horizontal circle* each of its two position components follows simple harmonic motion.

Section 2.3

In the case of a mass on a spring, the angular frequency of the oscillations can be predicted if m is known and k is known or can be deduced. The amplitude and phase of the oscillations can be found only if information about how the motion started is available.

Section 2.4

The model has one drawback when compared with reality—it predicts oscillations which continue with the same amplitude forever. In practice, the oscillations always die away and so the amplitude decreases. If resistive forces are taken into account in the model then *decaying oscillations* are predicted provided those forces are not too large. Once again, a fixed period of oscillations is predicted, but this period is greater than that predicted by a model which ignores the resistive forces.

Section 2.5

The *side-to-side oscillations of a pendulum* can be modelled by the differential equation

Section 3

$$\ddot{\theta} + \frac{g}{l}\sin \theta = 0$$

where θ is the angular displacement from the rest position at time t and l is the length of the pendulum. If the oscillations are small (say of amplitude less than $10°$) then, because $\sin \theta \approx \theta$ for such angles, the motion can be modelled as being simple harmonic.

Once again, this model ignores resistive forces. Inclusion of such forces in the model predicts decaying oscillations. It also points up a problem clockmakers faced which is that a pendulum clock is liable to run faster as it becomes due for winding. This problem was overcome by special shaping of the jaws holding the pendulum.

Section 3.1

The model of Section 2.5 can be applied to an *accelerometer*—a device which measures acceleration. Here oscillations are undesirable and resistive forces are introduced artificially.

Section 4

ANSWERS TO SELF-ASSESSMENT QUESTIONS

Figure 23

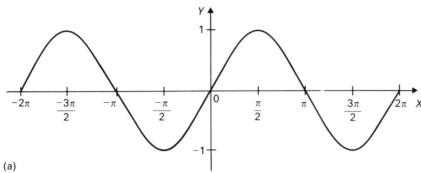

(a)

(a) The graph is shown in Figure 23(a). (i) $Y = 1$ at $X = -3\pi/2$, $\pi/2$; (ii) $Y = 0$ at $X = -2\pi, -\pi, 0, \pi, 2\pi$, (iii) $Y = -1$ at $X = -\pi/2, 3\pi/2$.

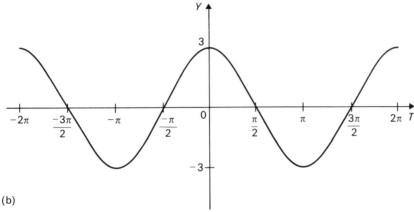

(b)

(b) The graph is shown in Figure 23(b). (i) $Y = 3$ at $X = -2\pi, 0$, 2π; (ii) $Y = 0$ at $X = -3\pi/2, -\pi/2, \pi/2, 3\pi/2$; (iii) $Y = -3$ at $X = -\pi, \pi$.

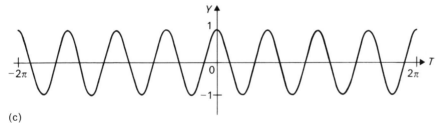

(c)

(c) The graph is shown in Figure 23(c). $Y = 0$ at $X = -15\pi/8$, $-13\pi/8, -11\pi/8, -9\pi/8, -7\pi/8, -5\pi/8, -3\pi/8, -\pi/8, \pi/8$, $3\pi/8, 5\pi/8, 7\pi/8, 9\pi/8, 11\pi/8, 13\pi/8, 15\pi/8$.

SAQ 2

(a) Two.

(b) $Y = P \sin 3X + Q \cos 3X$ where P and Q are arbitrary constants.

(c) Putting $Y = 0$ at $X = 0$ gives

$$0 = P \sin 0 + Q \cos 0$$

$$= 0 + Q$$

hence $Q = 0$.

To use the second condition, the general solution must be differentiated:

$$\frac{dY}{dX} = 3P \cos 3X - 3Q \sin 3X$$

Putting $dY/dX = 3$ at $X = 0$ gives

$$3 = 3P \cos 0 - 3Q \sin 0$$

$$= 3P - 0$$

Hence $P = 1$ and the particular solution is $Y = \sin 3X$.

SAQ 3

(a) (i) when $b^2 < 4ac$

   (ii) when $b^2 > 4ac$

(b) Here $A = 1$, $B = 2$ and $C = 5$. Hence $B^2 = 4$ and $4AC = 20$. Thus, $B < 4AC$ and the general solution is of the form (see *Handbook*)

$$Y = \exp\left(\frac{-BT}{2A}\right)\left[P \sin\left\{\frac{\sqrt{(4AC - B^2)}}{2A}T\right\}\right.$$

$$\left. + Q \cos\left\{\frac{\sqrt{(4AC - B^2)}}{2A}T\right\}\right]$$

where P and Q are arbitrary constants. Putting in the values for A, B and C gives

$$Y = \exp(-T)(P \sin 2T + Q \cos 2T)$$

SAQ 4

(a) $y = r\sin(\omega t + \varepsilon)$

Use the function of a function rule and put $u = \omega t + \varepsilon$

Then $y = r\sin u$ and $u = \omega t + \varepsilon$

$$\frac{dy}{du} = r\cos u \quad \frac{du}{dt} = \omega$$

$$\dot{y} = \frac{dy}{dt} = \frac{dy}{du} \times \frac{du}{dt}$$

$$= r\cos u \times \omega$$

$$= r\omega\cos(\omega t + \varepsilon)$$

Similarly,

$$\ddot{y} = -r\omega^2\sin(\omega t + \varepsilon)$$

(b) $\ddot{y} = -\omega^2\{r\sin(\omega t + \varepsilon)\}$

$$= -\omega^2 y.$$

Hence $\ddot{y} + \omega^2 y = 0$ and so $y = r\sin(\omega t + \varepsilon)$ is a solution.

SAQ 5

Here $P = 5$ and $Q = 12$.

Using equation (5),

$$R = \sqrt{(5^2 + 12^2)}$$

$$= \sqrt{169}$$

$$= 13$$

Using equation (6),

$$\tan\varepsilon = \frac{12}{5}$$

$$= 2.4$$

Hence

$$\varepsilon = 1.2 \text{ radians or } (\pi + 1.2) \text{ radians (to two significant figures)}$$

Both P and Q are positive, so both $\sin\varepsilon$ and $\cos\varepsilon$ must be positive. $\varepsilon = 1.2$ fulfils this condition; $\varepsilon = \pi + 1.2$ does not. Hence, $\varepsilon = 1.2$ and $Y = 13\sin(4T + 1.2)$.

SAQ 6

Figure 24 shows the completed sketch. Notice that, since the displacement is half what it was in Figure 5(c), the force is also half what it was in Figure 5(c).

SAQ 7

(a) 3

(b) 5

(c) 2.5

SAQ 8

(a) 2

(b) 3

(c) $\frac{1}{2}$

SAQ 9

(a) frequency $= 1/\tau$

$$= 1/5 \text{ Hz}$$

$$= 0.2 \text{ Hz}$$

angular frequency $= 2\pi/\tau$

$$= 2\pi/5 \text{ s}^{-1}$$

$$= 1.3 \text{ s}^{-1} \text{ (to two significant figures)}$$

(b) Periodic time $= 1/f$

$$= 1/10 \text{ s}$$

$$= 0.1 \text{ s}$$

angular frequency $= 2\pi f$

$$= 20\pi \text{ s}^{-1}$$

$$= 63 \text{ s}^{-1} \quad \text{(to two significant figures)}$$

SAQ 10

(a) $\pi/6$

(b) $\pi/3$

(c) -1

SAQ 11

(a) Amplitude $= 3 \text{ mm}$

phase $= 0$

angular frequency $= 0.2 \text{ s}^{-1}$

$$\text{frequency} = \frac{0.2}{2\pi} \text{ Hz}$$

$$= 0.032 \text{ Hz} \quad \text{(to two significant figures)}$$

$$\text{periodic time} = \frac{2\pi}{0.2} \text{ s}$$

$$= 10\pi \text{ s}$$

$$= 31 \text{ s} \quad \text{(to two significant figures)}$$

(b) Write this in the form (cf. SAQ 5)

$$Y = R\sin(2T + \varepsilon)$$

From equation (5),

$$R = \sqrt{(5^2 + 12^2)}$$

$$= 13$$

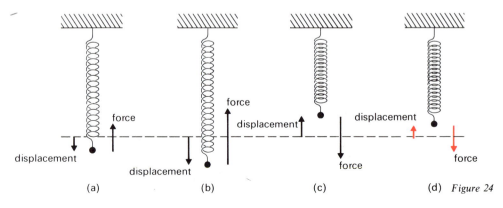

(a)   (b)   (c)   (d) *Figure 24*

From equation (6),

$$\tan \varepsilon = \frac{12}{5}$$

So $\varepsilon = 1.2$ radians (to two significant figures)

(This value was checked in SAQ 5, so there is no need to check it here.)

Hence amplitude $= 13\,\text{mm}$

phase $= 1.2\,\text{radians}$

angular frequency $= 2\,\text{s}^{-1}$

frequency $= \dfrac{2}{2\pi}\,\text{Hz}$

$\qquad = 0.32\,\text{Hz}$ (to two significant figures)

periodic time $= \dfrac{2\pi}{2}$

$\qquad = 3.1\,\text{s}$ (to two significant figures)

SAQ 12

Since the spring is unchanged, k still has the value $19.6\,\text{N}\,\text{m}^{-1}$. The periodic time τ is given by

$$\tau = 2\pi \sqrt{\frac{m}{k}}$$

$$= 2\pi \sqrt{\frac{0.4}{19.6}}\,\text{s}$$

$$= 0.90\,\text{s} \quad \text{(to two significant figures)}$$

SAQ 13

The general solution is

$$Y = P \sin 3T + Q \cos 3T$$

where P and Q are arbitrary constants. Since $Y = 0.01$ when $T = 0$,

$$0.01 = P \sin 0 + Q \cos 0$$

$$= 0 + Q$$

Hence

$$Q = 0.01.$$

Differentiating the general solution gives

$$\dot{Y} = 3P \cos 3T - 3Q \sin 3T$$

and putting $\dot{Y} = 0.03$ when $T = 0$ gives

$$0.03 = 3P \cos 0 - 3Q \sin 0$$

$$= 3P - 0$$

Hence $P = 0.01$ and the particular solution is

$$Y = 0.01 \sin 3T + 0.01 \cos 3T$$

To find the amplitude it is necessary to find R in the form of solution $Y = R \sin(3T + \varepsilon)$. It is not necessary to find ε since the question does not require it.

Using equation (5),

$$R = \sqrt{(0.01^2 + 0.01^2)}$$

$$= \sqrt{0.0002}$$

$$= 0.014 \quad \text{(to two significant figures)}$$

Hence the amplitude is $0.014\,\text{m}$. The periodic time is $2\pi/3\,\text{s} = 2.1\,\text{s}$ (to two significant figures).

SAQ 14

(a) The differential equation is

$$0.5\ddot{Y} + 18Y = 0$$

where $Y\,\text{m}$ is the extension at time T seconds. This can be rearranged to give

$$\ddot{Y} + 36Y = 0$$

and this has general solution

$$Y = P \sin 6T + Q \cos 6T$$

where P and Q are arbitrary constants.

(b) The initial conditions are $Y = -0.05$ at $T = 0$ and $\dot{Y} = 0.03$ at $T = 0$.

Putting $Y = -0.05$ and $T = 0$ in the general solution gives

$$-0.05 = P \sin 0 + Q \cos 0$$

hence $Q = -0.05$.

Differentiating the general solution gives

$$\dot{Y} = 6P \cos 6T - 6Q \sin 6T$$

Putting $\dot{Y} = 0.03$ and $T = 0$ gives

$$0.03 = 6P \cos 0 - 6Q \sin 0$$

$$= 6P - 0$$

Hence $P = 0.005$ and the particular solution is

$$Y = 0.05 \sin 6T - 0.005 \cos 6T$$

(c) To find the amplitude I have to write this solution in the form

$$Y = R \sin(6T + \varepsilon)$$

From equation (5)

$$R = \sqrt{(0.05^2 + 0.005^2)}$$

$$= 0.05 \quad \text{(to two significant figures)}$$

Hence the amplitude is $0.05\,\text{m}$ and the periodic time is $2\pi/6$ seconds $= 1.0$ seconds (to two significant figures).

SAQ 15

The periodic time is

$$= 2\pi \sqrt{\left(\frac{l}{g}\right)}$$

The length l of pendulum for a given periodic time τ

$$l = \frac{\tau^2 g}{4\pi^2}$$

For the present case, $\tau = 1\,\text{s}$ and $g = 9.8\,\text{m}\,\text{s}^{-2}$, so

$$l = \frac{1 \times 9.8}{4\pi^2}\,\text{m}$$

$$= 0.25\,\text{m} \quad \text{(to two significant figures)}$$

Because the approximation $\sin \theta \approx \theta$ is better for small angles and there is a chance that in the solution of differential equation (15) exactly as it stands there is no constant period, a clockmaker would want to make θ very small in order to obtain good time-keeping. In addition, this value for l has been calculated on the supposition of no resistance to motion. The clockmaker might want to make a correction to the length to allow for the presence of resistance in any real situation. He would also need to know the value of g in the location where the clock was to be used.

SAQ 16

The horizontal displacement, x, of a pendulum bob from its rest position is given by

$$x = l \sin \theta$$

So long as θ is small enough that $\sin \theta = \theta$ is an acceptable approximation

$$x = l\theta$$

$$\dot{x} = l\dot{\theta} \quad \text{and} \quad \ddot{x} = l\ddot{\theta}$$

In simple harmonic motion

$$\ddot{\theta} = -\omega^2 \theta$$

so

$$l\ddot{\theta} = -\omega^2 l\theta$$

and

$$\ddot{x} = -\omega^2 x$$

The horizontal motion is simple harmonic to the precision of the $\theta = \sin \theta$ approximation.

SAQ 17

SAQ 16 showed that the horizontal motion can be modelled as simple harmonic motion provided the displacement is small. Hence the displacement, X m, can be described by

$$X = P \sin \omega T + Q \cos \omega T$$

From the question, $X = 0.15$ and $\dot{X} = 0$ at $T = 0$. Using $X = 0.15$ at $T = 0$ gives

$$0.15 = P \sin 0 + Q \cos 0$$

so $0.15 = Q$

Differentiating the general solution gives

$$\dot{X} = \omega P \cos \omega T - \omega Q \sin \omega T$$

and using $\dot{X} = 0$ at $T = 0$ gives

$$0 = \omega P \cos 0 + \omega Q \sin 0$$

Hence $P = 0$ and the particular solution is

$$X = 0.15 \cos \omega T$$

Sufficient information is given in the question to enable ω to be found, since $\omega = \sqrt{(g/l)}$.

Hence

$$\omega = \sqrt{\frac{9.8}{1}}$$

$$= 3.1 \quad \text{(to two significant figures)}$$

and the required equation is

$$X = 0.15 \cos 3.1 T$$

SAQ 18

(a) There would be decaying oscillations provided

$$r^2 < 4m^2 l g$$

(b) The general solution is (see *Handbook*)

$$\theta = \exp\left(-\frac{rt}{2ml}\right)\left[p \sin\left\{\frac{\sqrt{(4m^2 l g - r^2)}}{2ml}t\right\}\right.$$

$$\left. + q \cos\left\{\frac{\sqrt{(4m^2 l g - r^2)}}{2ml}t\right\}\right]$$

where p and q are arbitrary constants.

SAQ 19

(a) Here $m = 0.1$ kg, $l = 1$ m, $g = 10$ m s^{-2} and $r = 0.1/0.5$ N s $= 0.2$ N s. So the differential equation is

$$0.1 \times 1\ddot{\theta} + 0.2\dot{\theta} + 0.1 \times 10\theta = 0$$

which is

$$0.1\ddot{\theta} + 0.2\dot{\theta} + \theta = 0$$

(b) $B^2 = 0.4$ and $4AC = 0.4$.

So $B^2 < 4AC$ and the solution is (see *Handbook*)

$$\theta = \exp\left(\frac{-0.2T}{0.2}\right)\left[P \sin\left\{\frac{\sqrt{(0.4 - 0.04)}}{0.2}T\right\}\right.$$

$$\left. + Q \cos\left\{\frac{\sqrt{(0.4 - 0.04)}}{0.2}T\right\}\right]$$

where P and Q are arbitrary constants. This simplifies to

$$\theta = \exp(-T)(P \sin 3T + Q \cos 3T)$$

(c) The initial conditions are $\theta = 0$, $\dot{\theta} = 0.3$ at $T = 0$.

Putting $\theta = 0$ at $T = 0$ into the general solution gives

$$0 = \exp 0(P \sin 0 + Q \cos 0)$$

$$= 1(0 + Q)$$

So $Q = 0$

Therefore, the solution is of the form

$$\theta = \exp(-T)P \sin 3T$$

To use the other initial condition, an expression for $\dot{\theta}$ is required. θ can be differentiated using the product rule:

$$\frac{d\theta}{dT} = G\frac{dH}{dT} + \frac{dG}{dT}H$$

where $G = \exp(-T)$, $\frac{dG}{dT} = -\exp(-T)$

$$H = P \sin 3T, \frac{dH}{dT} = 3P \cos 3T$$

Hence

$$\dot{\theta} = \exp(-T)3P \cos 3T - \exp(-T)P \sin 3T$$

Putting $\dot{\theta} = 0.3$ at $T = 0$ gives

$$0.3 = \exp 0 (3P \cos 0) - \exp 0 (P \sin 0)$$

$$= 1(3P \times 1) - 0$$

So $P = 0.1$ and the particular solution is

$$\theta = 0.1 \exp(-T) \sin 3T$$

(d) The graph is shown in Figure 25 overleaf. Note:

(i) Initial displacement $= 0$

(ii) Initial gradient represents initial velocity and so represents 0.3 rad s^{-1}

(iii) Period $= \dfrac{2\pi}{3}$ seconds

(iv) Exponential decay has time constant 1 second.

SAQ 20

The order is

(d) $r^2 < 4mk$ (b) $r^2 > 4mk$
(or possibly $r^2 = 4mk$)

(c) $r^2 < 4mk$

(e) $r^2 < 4mk$ (a) $r^2 > 4mk$

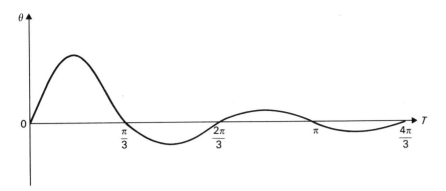

Figure 25

16. Review

CONTENTS

AIMS

The aims of this unit are:

1   To review, and take further, one course theme; the mathematical modelling of populations.

2   To increase your experience of mathematical modelling.

3   To provide opportunities for revision of some of the techniques you have met during the course.

OBJECTIVES

After reading this unit you should be able to discuss a number of mathematical models of population, including some models of two interacting populations.

STUDY GUIDE

As you will see from the Objectives, there is no new mathematics in this unit. Its purpose is to review and consolidate mathematical and modelling skills that you have met in the course and I hope you will find that it provides a good many opportunities for revision. However, the unit is by no means purely revision, as it will take a good deal further the discussion of one of the modelling themes—population. Nor does the unit provide opportunities to revise *all* the material in the course. It is intended to round off your studies in the course, provide a few pointers for the future and start you on your examination revision.

In this unit I shall not be using the convention that we have adopted so far in the course; that of using *capital* letters for numbers, and *lower case* letters for *dimensioned* quantities. I shall use capitals or lower case letters as I like—they may stand for either numbers or dimensioned quantities. This is to provide a link to the 'outside world': the convention is purely one adopted in *this* course and if you read other mathematics books, or even go on to further Open University courses involving mathematics, you will not find it used.

1 MODELLING POPULATIONS

In 1955 the first pair of collared doves was recorded breeding in Britain, but in twenty years the number of these birds has grown so rapidly that they are now to be found in virtually every part of the British Isles. They are now regarded as a pest and are no longer a protected species.

The collared dove was a natural immigrant; its spread in Britain followed a similar, almost equally rapid, spread throughout Europe. This is in contrast to many other examples of such population explosions, which often follow the introduction by man of a foreign species to a new habitat. The rabbit in Australia and the grey squirrel in Britain are examples of 'pests' that man has brought on himself in this way.

Another, and quite separate, contrast between the spread of the collared dove and that of the grey squirrel is that the collared dove does not seem to have affected the populations of any of our native species of birds. The spread of the grey squirrel resulted in the decline of the native red squirrel, which is now extinct in most of its former range and can only be found in a few places that the grey squirrel has not reached: the red and grey squirrel compete too closely to coexist. The collared dove seems similar enough to the wood pigeon, another 'pest', but apparently their populations do not significantly affect each other. The collared dove prefers a suburban habitat, the wood pigeon the country, but the significant differences are more probably in the choice of nest sites and in feeding habits.

Why do some species suddenly increase in numbers, while others stay constant, and yet others decline? Why does one or another species win out in competition? Why can some apparently similar species coexist, while others cannot? To give proper attention to such questions I would need to write a whole course, but one aspect of the scientific study of populations is the use of mathematical models and these are the subject of this unit. I shall be reviewing population models that you have met in the course and introducing some new ones. The unit is concerned with *modelling* and concentrating on one modelling theme will help with this.

It is not easy to make general statements as to the purpose of a whole class of models, even when they are all in the same area, and it is inappropriate to try. The purpose of mathematical models of populations in ecology might be: 'to predict the effect of man's behaviour on other species'; but to be able to do this one needs to understand the factors affecting natural species and the interactions between species, and models may be set up to assist in this understanding. The use of mathematical models of populations is not confined to ecology: they are, for example, used to predict human populations. They may also be set up for quite specific purposes—to predict the best way of exploiting a fishery, an example we have considered previously in TV8 and Unit 12, or to demonstrate the effect of a proposed method of control of an insect pest.

I cannot, in the space available, look at all the models of populations that I might wish to consider, so I must be selective. I can divide the models that I will consider into two sorts. I shall be looking at models to predict the size of a population at some future date: I shall look at theoretical models of the growth of a population and at how such theoretical models may need to be modified to correspond more accurately with available data. Secondly, I shall look at some models describing the interaction of two species—predator and prey. In this case I shall not be looking for *accuracy*, but only for the predictions of the models to fit the most general features of the available data. My purpose here will be to investigate the way in which the two species affect each other.

Throughout the unit I hope to pay closer attention to one part of the modelling cycle, the 'return to reality', than we have usually had leisure for in earlier units. I also hope to show you how the development of equations of increasing sophistication enables me to set up models to examine in progressively finer detail the way populations change.

I have been talking in this section as though populations should usually be expected to *change* in size. We are accustomed to expect change—growth— in human populations, but would you expect the population of a common bird, such as the great tit, to grow from year to year? My first model of such a population would, rather, be that the population remained constant. Even a population like this *does* change, however. As we noted in Population in *Modelling Themes*, there is a variation in size within each year. If I want to ignore this seasonal change in population, so as to look at the variation in size from year to year, I must do so by measuring 'population size' in a suitable way. The number of pairs of great tits breeding in a specified area is a measure of population size that is independent of such variations within a year. If I measure the population size in this way, how will my proposed model of constant population size stand up in practice? This is the first question that I shall consider.

2 PREDICTING THE UNPREDICTABLE

2.1 Herons and pied flycatchers

At the end of the introduction, I proposed a model of the way many natural populations change from year to year; that they do not. Let us compare this with some data on real populations. Figure 1 shows data on populations of two species of British birds. A glance at these graphs does not suggest that the populations are constant!

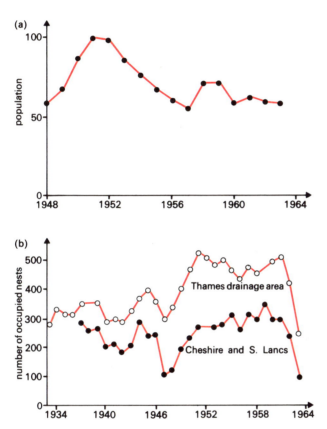

Table 1 Data for the population of pied flycatchers shown in Figure 1(a)

year	number of pairs
1948	58
1949	67
1950	87
1951	100
1952	98
1953	85
1954	76
1955	67
1956	60
1957	54
1958	71
1959	71
1960	58
1961	62
1962	59
1963	58

Figure 1   Population of two species of British birds, measured by the number of breeding pairs. (a) The population of the pied flycatcher in the Forest of Dean. (b) The population of the grey heron at two locations.

There is no clear trend in the way these populations change; they seem to rise and fall at random. It seems likely that the number of these species that can survive varies from year to year, depending, perhaps on the weather.

The three coldest winters in Britain between 1930 and 1965 (measured by the lowest average temperature from December to February) occurred in 1940, 1947 and 1963. These do correspond to the most dramatic falls in the heron population. (The pied flycatcher is a migrant, only visiting Britain in the summer, so there is no point in looking for the effect of these winters in this case.) Assuming that I cannot predict the weather, how can I predict future levels of these populations?

To model these populations, I will return to my original idea that the populations are constant. This is not tenable as it stands, but I can modify it to something more reasonable. I shall suppose that the populations vary at random about some fixed level and that over the years the population that can survive does not change, but that in any given year the population may deviate from this level, due to the fluctuations in weather, or whatever.

In predicting future population levels I can *not* hope to make a prediction of the form: 'the population of herons in the Thames area will be 400 in 1980' and have any confidence that I will be accurate. However, this is not an unusual situation when making predictions. If you *are* making a prediction of which you are unsure, it is helpful to admit it. I do not mean that you should phrase your prediction in the form: 'my best guess is that the population of herons will be about 400, but I am sure this is wrong'. I mean that you should seek to make a prediction of the form: 'I expect that the population of herons will lie between 200 and 600'. I shall seek to make a prediction of this type. Obviously, the narrower the range in which I predict that the population will be, the more informative is my prediction.

A simple way to make such a prediction would be just to choose the largest and smallest observed value of the population as the bounds within which you expect it to fall. On this basis, I should predict that the population of the pied flycatcher will always be between 54 and 100. Using this approach, I do not know if the observed data cover all the possible values that the population may take or, on the other hand, if one of these values is very unusual and outside the typical range of variation. I shall therefore try a more elaborate approach.

I shall suppose that the frequency of occurrence of the various population sizes is given by a *Gaussian distribution* (Unit 1, Section 5).

What parameters do I need to specify a Gaussian distribution?

Its mean and its standard deviation.

I shall choose the mean and standard deviation for my model of the distribution of population sizes to be equal to the mean and standard deviation of the data.

How do you calculate the mean and standard deviation of a list of numbers?

The mean is the sum of the numbers divided by the number of numbers. To find the standard deviation you must: (a) calculate the difference between each number and the mean; (b) square these results; (c) calculate the mean of these squares; and (d) take the square root of the result of (c). (Unit 1, Section 6.)

The data on the pied flycatcher population corresponding to the graph in Figure 1(a) are given in Table 1. The mean population size is 74.3 and its standard deviation is 14.6.

SAQ 1 (Unit 1, Section 5)

Use the model that I have described to give bounds within which you expect the pied flycatcher population to lie.

SAQ 1

In the solution to SAQ 1, I suggested that you can be confident that the pied flycatcher population will lie within three standard deviations of the mean. This gives rather a large interval within which I am predicting that the population will lie. I can give a smaller interval if I am prepared to be less certain of my prediction. I might look to within two standard deviations of the mean rather than three. This gives me a smaller interval, but just how sure can I be that the population will fall inside this interval (or within three standard deviations, for that matter)?

I can answer this question by using known information about the Gaussian distribution. The proportion of the total area under the Gaussian curve that falls within the specified bounds (see Figure 2) can be found in statistical tables. The area between plus and minus three standard deviations of the mean is about 0.997 of the total area (remember that the *total* area under the curve is 1). On this model, I therefore expect the value of the population to lie in the specified range (i.e. 31 to 118) 99.7% of the time. If I predict that the population will lie within *two* standard deviations (i.e. between 45 and 104), then I can be less confident. I would only expect the population to lie in this range about 95% of the time.

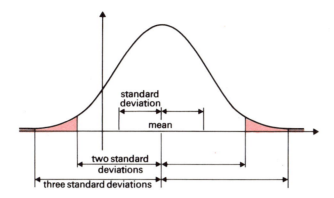

Figure 2 The graph of a Gaussian distribution showing the area not within three standard deviations of the mean and the area not within two standard deviations of the mean. These shaded areas are less than 0.3% and 4.6%, respectively, of the total area under the graph.

Remember that these predictions depend on the suppositions I have made. If these are not accurate, even my most confident prediction could be wrong.

What is the main supposition I made?

That the distribution of population sizes is *Gaussian* (rather than any other sort of distribution).

Can you suggest any way of testing this supposition against reality?

The natural test to perform is to make a histogram of the data in Table 1 and see if it approximates to a Gaussian distribution. However, I shall not do this, as there are simply not enough data (only sixteen items) to expect such a histogram to exhibit the Gaussian shape (or, indeed, any clear shape).

To summarize so far, I started with the idea that many populations do not change in size and modified it to a picture of a population fluctuating 'at random' about a fixed mean. That is to say, I have supposed that there is no *trend* in the population size; the variations are unpredictable. My supposition that the distribution of population sizes is Gaussian has enabled me to predict the unpredictable—to some extent at least. I want next to look for the possibility of trends in such a fluctuating population.

Study comment

The branch of mathematics that deal with models to cope with the problems of 'randomness' or 'unpredictability' is known as *statistics*. We have only been able to give you a very small taste of this subject in this course. For instance, we have only looked at one of the many important forms of distribution that are used in statistical models—the Gaussian distribution.

2.2 Homo sapiens and great tits

Figure 3 gives data on the populations of two other species native to Britain (although the particular great tit population considered is from a location in Holland). Do you think that there is a trend in the size of these populations, or do you think that a random variation about a fixed mean is a better description?

For the data in Figure 3(a) on the human population, I hope the answer is obvious! The question is a much more interesting one for the great tit population in Figure 3(b), and I shall look at that later.

SAQ 2 (Unit 2)

Set up a model to predict the human population in 1911 from the data in Table 2 (which is plotted in Figure 3(a)).

SAQ 2

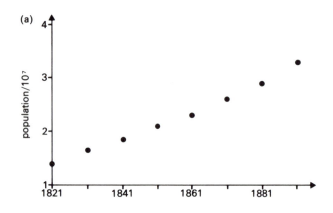

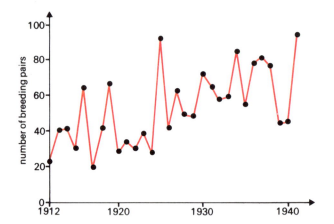

Table 2 Population of Great Britain, 1821–91, to the nearest million

year	population
1821	14 000 000
1831	16 500 000
1841	18 500 000
1851	21 000 000
1861	23 000 000
1871	26 000 000
1881	29 500 000
1891	33 000 000

Figure 3   Do these populations show a trend? (a) The human population of Britain between 1821 and 1891. (b) The number of breeding pairs of the great tit at Orabje Nassau's Oord, near Wageningen, Holland.

Would you be happy to use a linear model, as set up in the solution to SAQ 2, to predict the human population in 2011?

I hope you would not: recognizing that a straight line fits the data reasonably well, and then extrapolating, gives a reasonably accurate prediction if the period of time over which you are extrapolating is short. To extrapolate over a longer period—in this case longer than the time span of the existing data—is likely to be inaccurate. It may be the best (or the only) method that you can think of for making a prediction, in which case you will have to use it, but do not be confident of its accuracy if you do.

In this example, I can think of an alternative way of extrapolating the data which also has some 'theoretical' justification. For a longer term prediction, I prefer to fit a (shallow) exponential curve.

I shall turn now to the question of deciding whether or not there is a trend in the population of great tits (Figure 3(b)). One way to predict future population sizes is to imagine the population varying at random about the mean population size, which is 51.4. Figure 4(a) illustrates that the population does vary to quite a large extent about this mean. The extent of this variation is measured by the standard deviation of the data—20.1.

Figure 3(b) suggests to me that there is a certain upward trend in the population size. The simplest model of an upward trend that I can use is a linear model and a possible linear model for predicting this population is shown in Figure 4(b). I can imagine the variation in population size as a random variation superimposed on this *linear* trend, rather than simply as a random variation about a fixed average value.

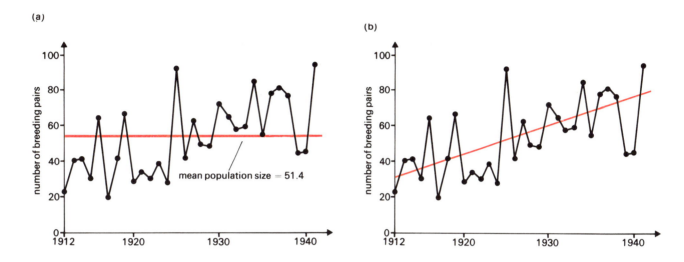

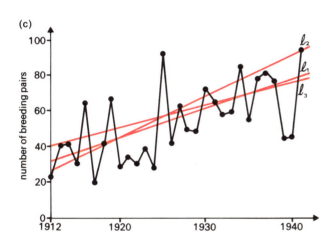

Figure 4   Possible models for predicting the great tit population of Figure 3(b): (a) 'random' variation about the mean; (b) a possible linear increase; (c) three other possible linear models.

How can I decide if this is a better model? Does it fit the data more accurately?

To do this I, first, calculate the amounts by which the observed values of the data differ from those suggested by the straight line graph. Then calculate the square root of the mean of the squares of these 'errors'. This quantity measures the extent of the variation of the data about the increasing straight line in the same sort of way that the standard deviation measures its variation about the horizontal straight line—the fixed mean—of Figure 4(a). If this quantity is smaller than the standard deviation then I have a basis for claiming that the increasing straight line is a better fit to the data than the horizontal line of Figure 4(a).

There is more than one apparently reasonably way of fitting an increasing straight line to the data (see for example Figure 4(c)). I can use the same criterion to decide which is preferable.

SAQ 3

Figure 4(c) shows three possible straight lines l_1 l_2 and l_3 that might be used to predict the population of great tits. The equation of l_1 is $p = 30 + \frac{3}{2}t$; that of l_2 is $p = 25 + 2t$; and that of l_3 is $p = 40 + t$, where p is the population at time t years after 1912. The variation of the data about l_1 (measured as suggested in the text) is 16.2; its variation about l_2 is 17.3; and that about l_3 is 17.0. The standard deviation of the data is 20.1.

(a) What model would you use to predict this population?

(b) If you had to give a single number as the predicted population for 1978, what would it be? How confident are you of this prediction?

(c) Is there any other form in which you would prefer to give your prediction of the population in 1978? If so, give a prediction in this form.

SAQ 4

Data on a population of herons are given in Figure 1(b). Would you expect that population to increase, decrease, or stay the same, for the following year (i.e. 1964)?

2.3 Summary

In this section I have looked briefly at some ways of coping with 'randomness' and also at extrapolating data by fitting straight lines to the data. I have only been able to give you an introduction to these ideas; there is a great deal more to both of these subjects.

3 POPULATION INCREASE

In Section 2.2, I considered linear models for predicting the growth of two populations. Linear models are convenient since they are mathematically simple. If I am to predict the human population of Britain in 2011 from the data in Table 2, choosing a model that fits the data reasonably well, and is mathematically convenient, may not be enough to yield an accurate prediction. I can produce other models (even other linear models) that fit that data equally well and produce quite different predictions when extrapolated so far ahead. In many branches of science predictions are made far ahead and with the expectation of accuracy. This confidence is based on having an underlying *theory*. I want now to consider a theory of population growth.

In this theory I shall set aside the element of random fluctuation that I considered in the previous section. (I would need statistical models to take that into account.) When I come to compare my theoretical models with data, then this random element will inevitably be present in the data. The theory will help you revise material in Units 3, 8 and 12.

3.1 Exponential growth

Imagine a very simple organism which reproduces by dividing into two. Suppose each organism does this after a time period τ.

How many individuals will there be after a time t, if there is one to start with?

There will be $2^{t/\tau}$. See the following text.

The numbers will grow as given in Table 3.

Table 3

time	0	τ	2τ	3τ	4τ	5τ	
number	1	2	$4 = 2^2$	$8 = 2^3$	2^4	2^5	etc.

It would be natural to describe such population growth by a discrete model, and say that the population after N time periods is 2^N. I want, however, to concentrate on continuous models. The population is 2^N after N periods: that is, at time $t = N\tau$. Thus $N = t/\tau$ and the population, P, at time, t, is

$$P = 2^{t/\tau}.$$

Allowing t to take a continuous range of values, corresponds to a smooth interpolation between the values in the table above.

This population is growing *exponentially* (Unit 8). It is more usual to express exponential growth in terms of the exponential function and, for some constant r, to write

$$P = e^{rt} \tag{1}$$

If I do this, what is r in terms of τ?

$$r = \frac{1}{\tau}\ln 2$$

13

Exponential growth is to be expected in any situation where the rate of increase is proportional to the existing size of the population (or whatever). This statement can be formulated mathematically as

$$\frac{dy}{dt} \propto y$$

or

$$\frac{dy}{dt} = ry \qquad (2)$$

What is the general solution of the differential equation (2)?

$y = ae^{rt}$, where a is the arbitrary constant.

What does a represent here?

The value of y when $t = 0$.

As the first thought about the way a population will change, the idea that the rate of change is proportional to the existing population seems reasonable. It seems natural to expect that the number of births will be proportional to the existing population; and so will the number of deaths (unless the age structure of the population is very variable). Even emigration might reasonably be expected to be proportional to the existing population size; if immigration were a significant factor, there would seem no reason to expect this to be proportional to the existing population.

The parameter r in the differential equation $dy/dt = ry$ is called the *instantaneous fractional growth rate*. Notice that this is different from the simple 'instantaneous rate of change' of the population. The 'instantaneous rate of change' of y (with time) is dy/dt: the 'instantaneous fractional growth rate' is

$$\frac{1}{y}\frac{dy}{dt}$$

Exponential growth arises from a supposition of *constant fractional growth rate* (a supposition that the rate of change is in a fixed proportion to the existing population).

What prediction would arise from a supposition of constant *rate of change*?

A prediction of *linear growth*.

SAQ 5 (Unit 8)

(a) What are the dimensions of the instantaneous fractional growth rate?

(b) If the population of an organism trebles every year, what is the instantaneous fractional growth rate of this population?

(c) If a population has an instantaneous fractional growth rate, r, what is the time period in which it will double its numbers?

If τ is one year then the population whose growth is shown in Table 3 doubles each year. This does *not* mean that the instantaneous fractional growth rate is then two per year. The instantaneous fractional growth rate is in fact ln 2 per year: 'two per year' is an *average* fractional growth rate.

A model of exponential growth will predict that the population will grow in a fixed proportion in any one year, but this proportion is *not* the same as the instantaneous fractional growth rate. It is a fractional growth rate averaged

14

over a year. I shall explain this. Suppose the population p at time t years is given by

$$p = A \exp rt$$

(Remember my warning in the Study Guide: I am *not* using our previous course convention of capital letters and small letters for non-dimensioned and dimensioned quantities.)

One year later the population is

$$A \exp r(t + 1)$$

In that time the population has grown in the proportion

$$\frac{A \exp r(t + 1)}{A \exp rt} = \exp[r(t + 1) - rt]$$

$$= \exp r$$

This result is independent of t—in any year the population grows in the same proportion. But $\exp r$ is not equal to r. If the instantaneous fractional growth rate is r per year, the population grows each year by the factor $\exp r$, or e^r.

I have now developed a 'theoretical' model of population growth. How does this model of exponential population growth accord with reality? Do real populations grow exponentially?

Tables 4 and 5 give authentic data on two populations. How can I decide whether exponential growth is a reasonable model of these population explosions?

Table 4 British population of collared doves in various years

year	1955	1956	1957	1958	1959	1960	1961	1962	1963	1964
population	4	16	45	100	205	675	1900	4650	10 200	18 855

Table 5 The human population of the world between 1930 and 1970, in millions

year	1930	1940	1950	1960	1970
population	2070	2295	2485	2982	3635

The characteristic of population growth that I will look for is that the increase in any period of time is in a *fixed proportion* to the existing population.

I will look at the collared dove population for which data are given in Table 4. From 1955 to 1956 this population increased by $16 - 4 = 12$, which means that the annual increase was $12/4 \times 100\% = 300\%$. Calculating similarly the percentage increase for the various years gives the numbers in Table 6.

Although this annual percentage increase is by no means exactly constant, I can picture these numbers as varying randomly about some fixed value that they *should* have (but for weather variations, and so on). I feel that a model of exponential population growth fits these data reasonably well. At least it would provide some sort of basis for predicting numbers over the next few years.

Table 6 Annual increase in the collared dove population as a percentage of the previous population

year	% increase
1955–56	300
1956–57	181
1957–58	122
1958–59	105
1959–60	229
1960–61	181
1961–62	145
1962–63	119
1963–64	85

SAQ 6

Comment on the appropriateness of a model of exponential growth for describing and predicting the human population of the world for which data are given in Table 5.

SAQ 6

Although the data on neither the human nor the collared dove populations fit precisely a model of exponential growth, such models do supply a convenient framework for discussing these populations. They are certainly closer to reality than, for example, a model of linear population increase. You will notice that the collared dove population has a much greater fractional growth rate than the human population: collared doves reproduce faster than humans!

Having decided that an exponential model is suitable for estimating the future population levels of collared doves, a convenient technique when trying to fit the model to data is to plot the data on log–linear graph paper.

Why is this?

Because you now want to fit a *straight line* to the data points.

If you do not have any log–linear graph paper, the same effect can be obtained by plotting the *logarithms* of the data against time on ordinary graph paper. Either way you are drawing a graph of a new variable

$$Z = \ln P$$

against time. If P is increasing exponentially (that is, if $P = Ae^{rt}$) then

$$Z = \ln P = \ln A + rt$$

Thus the graph of Z against t is linear. Not only is it easier to try to fit a straight line to data, it is also possible to decide a value for the instantaneous fractional growth rate r from this approach—you use the gradient of your straight line.

SAQ 7 (Unit 8)

SAQ 7

(a) By plotting the logarithms of the data in Table 4, find an exponential model to predict the population of collared doves in 1966 and 1977.

(b) Have you any comments on these predictions?

If you choose to extrapolate from given data by fitting a straight line to the logarithms of the data, you should be clear that you are making a modelling supposition, that the data are growing (or decaying) *exponentially*. You can persuade yourself that a straight line is a reasonable fit to data in many cases; and the choice between fitting a straight line to the data or to their logarithms is the choice between a linear or an exponential model. With *population* growth, there is always an underlying expectation that it will be exponential.

Having decided on an underlying theory that populations change exponentially, it is arguable that it is always sensible to study the logarithms of population size. Had I looked at graphs of the logarithms of population size in Section 2, then I should have been looking at *relative*, rather than absolute, variations in numbers. On a graph of the actual population size a change from 200 to 1200 shows up as equally significant as a change from 100 200 to 101 200. In the first instance, however, the population has increased by a factor of six; in the second by 1%. Such relative variations are often of greater interest and will show up on a plot of logarithms. Figure 5 shows an example of the effect of looking at logarithms.

In this section I have noted that, under certain circumstances at least, natural populations may increase in a way that may be described by exponential growth. Such a population increase may, for example, occur when a species is introduced into a favourable habitat, or when it spreads naturally, as did the collared dove.

Although the population growth of my two examples, human beings and collared doves, could be described reasonably well by an exponential model, the model was not a perfect fit with the data in either case. I now

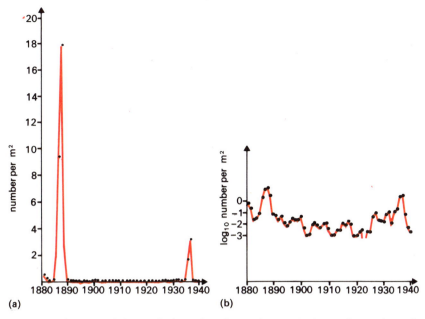

Figure 5   The effect of plotting the logarithm of a population. (a) The population of a moth (Dendrolimus) at Letzinger, Germany. (b) The logarithm of the population shows that Figure 5 (a) conceals a good deal of relative variation in population size.

want to look at possible modifications to this basic model that can improve the fit with the available data. Each example needs a different approach.

3.2   'Exponential-plus' growth: homo sapiens

The calculations in the solution to SAQ 6 suggest that the fractional growth rate of the human population of the world may itself be increasing. This is confirmed by the data in Table 1 of Population in *Modelling Themes* (reproduced here as Table 7), which shows that the time needed for the population to double has decreased, and is still doing so.

Table 7 World population and doubling times

date	estimated world population	time required for population to double
8000 BC	5 million	1500 years
AD 1650	500 million	200 years
AD 1850	1000 million	80 years
AD 1930	2000 million	45 years
AD 1975	4000 million	35 years*

* Computed doubling time around 1972.

Figure 5 in Population suggest a possible reason for this increase in the fractional growth rate. The graphs for Sri Lanka and Mexico show that the annual birth rate in those countries has tended to remain steady (at about 4%) even when the death rate has dropped dramatically. This suggests that human populations may tend to keep a constant birth rate, even when the death rate falls due to the introduction of improvements in agriculture, hygiene and medicine, and the spread of these improvements throughout the world. It is true that the graph in Figure 5 for Sweden shows the birth rate as well as the death rate decreasing, but this seems to be a characteristic only of developed countries in the very recent past. The effect of this on the birth rate of the world as a whole is so far only slight. (In 1972 the world's birth rate was estimated to be 3.3%.) If I suppose that the birth rate has been steady and the death rate has declined, the fractional growth rate of the population will have been increasing.

I shall now build a mathematical model of the world's population on the basis of the suppositions that the birth rate is constant, at 4% per year and, for convenience, that the death rate has declined *linearly*.

Therefore I shall suppose that the death rate, D per year, is given by the equation

$$D = 0.04 - KT$$

At $T = 0$, this equation gives $D = 0.04$, the value of the birth rate. So T is the number of years measured from some hypothetical point in the past when the birth and death rates were in equilibrium. In this equation K is a constant (which will be small and positive). I will only use the model for values of T such that

$$0 \leq T \leq \frac{0.04}{K}$$

as I do not want to suppose that the death rate exceeded the birth rate for $T < 0$, or that the death rate will become negative.

SAQ 8

Use these suppositions to set up a differential equation for the world population P at time T years.

SAQ 9 (Unit 12, Section 5)

Solve the differential equation for the world population P at time T years set up in SAQ 8.

I now have the following model of the growth of the world's human population, which might be called an 'exponential-plus' model, as it predicts growth even faster than an exponential. Using the solutions to SAQ 8 and SAQ 9, the population P at time T years is given by the equation

$$P = C \exp \frac{1}{2} K T^2 \qquad \left(0 \leq T \leq \frac{0.04}{K}\right) \qquad (3)$$

The time, T years is measured from the hypothetical moment in the past when the death rate started to fall.

I want now to examine how this model corresponds with the data on the human population in Table 7. To do this I will first choose a new time origin, at AD 1850, since I am measuring time from the hypothetical moment at which the death rate began to fall, which I do not know. Suppose this unknown moment is A years before 1850, and let T years from the unknown origin correspond to U years measured from 1850. Then $T = U + A$.

Substituting this in equation (3) gives

$$P = C \exp \frac{K}{2} (U + A)^2 \qquad (4)$$

How can I compare this model with the data?

In Section 3.1 I observed that it is more convenient when seeking to fit an exponential to data to look at the logarithms of the data. If I look at the logarithms, then I am in effect looking at a new variable $Q = \ln P$. From equation (4) and using the properties of logarithms and exponentials Q is given by

$$Q = \ln P = \ln C + \frac{K}{2} (U + A)^2 \qquad (5)$$

This is a quadratic function in U and it will be easier to try to fit equation (5) to the data than equation (4). Because of the nature of the data, it is still more convenient to look at $Z = \ln(P/P_0)$ where P_0 is the population in 1850. As you can see from Table 8, this makes the numbers simpler.

Table 8

year	8000 BC	AD 1650	AD 1850	AD 1930	AD 1975
U	-9850	-200	0	80	125
P/P_0	$\dfrac{1}{200}$	$\dfrac{1}{2}$	1	2	4
$Z = \ln(P/P_0)$	$-\ln 200$	$-\ln 2$	0	$\ln 2$	$\ln 4$

Now

$$Z = \ln(P/P_0)$$
$$= \ln P - \ln P_0$$
$$= Q - \ln P_0$$
$$= \ln C - \ln P_0 + \frac{1}{2}K(U + A)^2$$

For simplicity I will write this as

$$Z = S + \frac{1}{2}K(U + A)^2 \tag{6}$$

where S is a constant. I am at liberty to choose each of the parameters S, K and A, so as best to fit the data. There are three unknown parameters here, so I would expect to be able to choose them so as to fit three data points accurately—but I have five data points. I could choose to try to draw a parabola so as to fit all five data points as well as possible. Instead, however, I shall choose to fit accurately the three most recent data points. I would hope that this would give the best basis for predicting the immediate future and it will be mathematically simpler. It is not going to be easy to fit a parabola to data points by eye.

Putting the data for $U = 0, 80$, and 125 into equation (6) gives the equations

$$\left.\begin{aligned}
0 &= S + \frac{1}{2}KA^2 \\
\ln 2 &= S + \frac{1}{2}K(80 + A)^2 \\
\ln 4 &= S + \frac{1}{2}K(125 + A)^2
\end{aligned}\right\} \tag{7}$$

To solve these equations I first eliminate S by subtracting the first equation from each of the others. This gives

$$\left.\begin{aligned}
\ln 2 &= \frac{1}{2}K[(80 + A)^2 - A^2] \\[2mm]
\text{and}\qquad \ln 4 &= \frac{1}{2}K[(125 + A)^2 - A^2]
\end{aligned}\right\} \tag{8}$$

I can now eliminate K by division.

$$\frac{(80 + A)^2 - A^2}{(125 + A)^2 - A^2} = \frac{\ln 2}{\ln 4}$$

$$= \frac{\ln 2}{2\ln 2} = \frac{1}{2}$$

Solving for A is now a matter of algebra of the sort you met in Unit 3. I can find K and then S, by substituting back into equations (8) and (7). I shall omit the details, leaving you to fill them in *if you wish*. Working to two significant figures, I get: $K = 1.1 \times 10^{-4}$; $A = 40$; and $S = -8.6 \times 10^{-2}$.

Since $A = 40$, this model suggests that the death rate started to drop forty years before 1850, that is, in 1810. Therefore the model cannot be taken back before 1810 and cannot be a reasonable fit to the earlier data points. However, as it fits recent data accurately it is likely to form a good basis for short or medium-term predictions. For longer term predictions I should think it better to look for a parabola that fits the last four data points (including 1650) as well as possible.

With these values of K, A, and S I have a model of the world's human population that I might use for predicting population levels. This is

$$\ln(P/P_0) = S + \frac{1}{2}K(U + A)^2 \qquad 0 \le U + A \le \frac{0.04}{K} \qquad (9)$$

I will ask you to investigate the predictions of this model in SAQ 10.

SAQ 10

(Work to two significant figures throughout.)

(a) What population level does this model predict for 2010?

(b) What was the instantaneous fractional growth rate of the world population in 1975 according to this model?

(c) Imagine that in 1975 you had set up an ordinary exponential model based on the supposition that the instantaneous fractional growth rate remained constant at the level that you found in part (b). What doubling time would this model predict?

(d) According to the 'exponential-plus' model of equation (9), for how long will the *fractional growth rate* of the human population go on increasing?

The solution to SAQ 10 shows that, according to this model, the *rate of increase* of the world's population will go on increasing until about 2170. The prediction of the model for the population in that year is roughly 1.5 million million, over 300 times its present level. The population is predicted to double eight more times in the next two centuries: in the last century it doubled only twice. This is in accord with the prediction of the model that the doubling time continually decreases.

This model is a good fit with recent data and looks likely to make accurate short-term predictions, but can a population really go on increasing in the sort of unlimited way predicted by this model, or even by the simple exponential model? I shall look at this question now, in relation to the other 'exploding' population—that of collared doves.

3.3 Limits to growth: are collared doves logistic?

In the solution to SAQ 7, I predicted the British population of collared doves in 1977 by supposing exponential growth. That model predicts a population of about 1500 million birds in 1977. If there really were that many of them around, they would be about a hundred times as common as sparrows. Although the collared dove is now a reasonably common bird all over Great Britain, it is not quite so common as that! The population has continued to grow since 1964, but not so quickly as an exponential model predicts.

This is hardly a surprise. After all, the exponential model predicts a collared dove population of 1000 million this year; 10 000 million in about three years time; and 100 000 million in another three years. Either we end up knee-deep in collared doves or the population growth must slow down.

Why should this happen?

A natural supposition to make is that the environment can only support a certain number of these birds. The most obvious factor likely to limit the population is the available food supply. This surely places an upper limit on the number of the birds that can survive in Great Britain, although this may not be the crucial factor. Research has shown that, for some species, other factors cause a reduction in the growth rate before the population reaches the level where all the available food is being used.

Whatever the cause of this limitation, I want now to consider a model based on the supposition that there is a maximum population, call it M, of a species that the environment will support. I would like a model that predicts exponential growth at low populations levels, but one that also predicts that the fractional growth rate is zero when the population is equal to M. This suggests that I want a model described by the equation

$$\frac{1}{P}\frac{dP}{dt} = f(P) \tag{10}$$

where P is the population at time t, and $f(P)$ is some function of P that decreases as P increases and has reached zero when $P = M$. There are plenty of functions which have this property that I could choose for $f(P)$, but I have no basis for choosing any *particular* one.

What is the *simplest* function $f(P)$ that I could choose?

A *linearly* decreasing function; say

$$f(P) = a\left(1 - \frac{P}{M}\right) \quad \text{(where } a > 0\text{)}$$

Putting this form of $f(P)$ into equation (10) gives the *logistic equation*

$$\frac{1}{P}\frac{dP}{dt} = a\left(1 - \frac{P}{M}\right) \tag{11}$$

This equation is quite commonly used for describing 'growth with limits' (not only in the context of population modelling): you met it in Unit 12. I shall leave the solution to you, as revision.

SAQ 11 (Unit 12)

SAQ 11

Use the substitution

$$Q = \frac{M}{2} - P$$

(where Q is the new variable) to solve the logistic equation (11). Find the particular solution satisfying the initial condition $P = P_0$ when $t = 0$.

The solution to SAQ 11 shows that the logistic equation (11) has the solution

$$P = \frac{M}{1 + \left(\frac{M}{P_0} - 1\right)e^{-at}} \tag{12}$$

where P_0 is the population at $t = 0$. (This equation looks different from the solution to the logistic equation obtained in Unit 12, but the difference is just a matter of algebraic rearrangement.) I shall check that this solution has the sort of behaviour I am expecting.

What do I expect to happen to P as t becomes large?

I expect P to tend to M, the limiting population.

I must therefore check that equation (12) does imply this behavior. As t becomes large, e^{-at} becomes very small—remember that a is positive. In the limit, e^{-at} is zero and $(M/P_0 - 1)e^{-at}$ is zero, so P, as given by equation (12), tends to M as t becomes large.

Thus equation (12) implies the behaviour I expect for large values of t; the population is predicted to grow gradually up to the limiting population size M. It is a little harder to check what equation (12) predicts will happen initially, for small values of t. I will look at this question for the case when the initial population P_0 is much smaller than the limiting population M.

What would you expect to happen in this case?

Since P is well below M, I would not expect the limits to growth to have any significant effect and so I expect exponential growth.

Let us see if equation (12) predicts this. I will use my assumption that P_0/M is small to obtain a good approximation for P from equation (12).

$$P = \frac{M}{1 + \left(\dfrac{M}{P_0} - 1\right)e^{-at}}$$

$$= \frac{Me^{at}}{e^{at} + \dfrac{M}{P_0} - 1}$$

$$= \frac{P_0 e^{at}}{1 + \dfrac{P_0}{M}(e^{at} - 1)}.$$

If I confine my attention to small values of t, then e^{at} is not large. Since P_0/M is small, the term $P_0(e^{at} - 1)/M$ is then also small. For small values of t I can therefore neglect this term in the denominator and this gives

$$P \approx P_0 e^{at}$$

This is the usual equation for exponential growth with initial population P_0. So long as the initial population, P_0, is a good deal smaller than the limiting population, M, the logistic equation predicts that the early stages of population growth will be exponential, at an instantaneous fractional growth rate a.

Figure 6(a) shows the graph of the logistic curve described by equation (12). I have drawn the curve for negative as well as positive values of t. Choosing a solution to fit a particular value of the initial population, P_0, corresponds to shifting the curve to left or right (Figures 6(b) and 6(c)).

SAQ 12

Suppose you have some recent data on a population to which you can fit a logistic curve reasonably well: for example, that in Figure 7. Would it be reasonable to extrapolate backwards, to estimate past populations?

SAQ 12

SAQ 13

In Figure 6(a) I have shown that the logistic curve has a point of maximum slope with $P = M/2$. Show that this is correct. (There is no need to use equation (12) here; find the slope of the logistic curve as a function of P from equation (11).)

SAQ 13

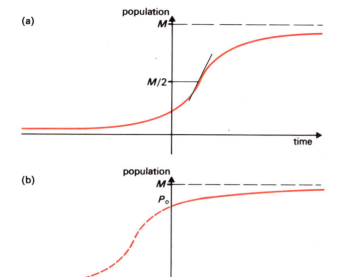

(a)

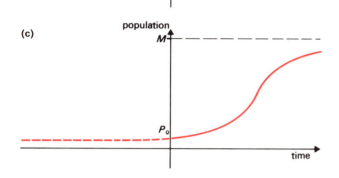

(b)

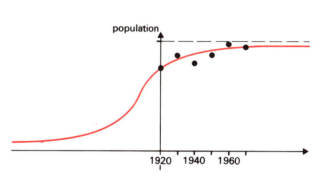

(c)

Figure 6 (a) The logistic curve; a popular model for describing 'growth with limits'. (b) and (c) show the different solutions to the differential equation (11), corresponding to different initial values of the population (P_0).

Figure 7   A logistic curve fitted to some recent population data. Is it reasonable to use this curve to estimate past populations, such as that in 1900?

So far in this section I have been extending my 'theory' of population growth. I have modified my model of exponential growth because I expect that, in the long term at least, the growth of a population will be limited: available food or some other factors will prevent it growing past some fixed level. The logistic curve is my new 'theory', revised to take this into account. Let me see now how this stands up in practice.

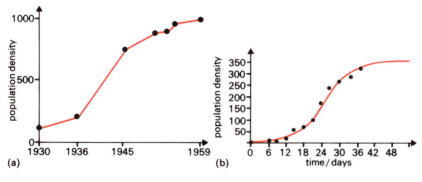

(a)   (b)

Figure 8   Examples of logistic growth of populations. (a) Population density of the grey squirrel (individuals per 100 km²) at selected sites in Great Britain. (b) The results of a laboratory experiment on a population of fruit flies.

It is not difficult to find examples of population growth exhibiting the general shape of the logistic curve. Figure 8(a) shows an example, the growth of the grey squirrel population in Britain. This shows an initial rapid growth, exponential in appearance, followed by a levelling off as the population reaches a limiting level. Clearly, something limited the growth of the grey squirrel population. There are also some well-known laboratory experiments on a species of fruit fly (*Drosophilia*) that produce population growth curves that are logistic in shape (Figure 8(b)). This information certainly supports a preference for the logistic model rather than the simple exponential model. Let me see if I can obtain a detailed fit with some specific data. I will look at the data on collared doves in Tables 4 and 6.

To compare the data with the model I have plotted (Figure 9) the annual growth divided by the population, against the population. I want to compare this with the supposition made in setting up in equation (11), that the growth rate $(1/P)/(dP/dT)$ declines linearly with population. I have had to plot the average, rather than the instantaneous, fractional growth rate as this is what I know. It should, however, be sufficient to indicate whether the model is reasonable. (You should remember from Unit 8 that these two rates are related.)

The data points in Figure 9 do not, as a whole, suggest a clear trend. However, the figures for the later dates are more likely to form a reliable basis for making predictions. As they also refer to larger populations they are the more likely to be reliable, in that they will be less affected by random elements. I shall concentrate therefore on the figures from 1959 onwards. For these figures, there is a clear downward trend in the fractional growth rate. If I model this trend with a straight line, in accordance with the logistic model, then I can obtain an estimate of the equilibrium population from this data.

SAQ 14

SAQ 14

Suppose that the fractional growth rate of collared doves declines linearly with population. Hence estimate from the data in Figure 9 the maximum population of collared doves in Britain.

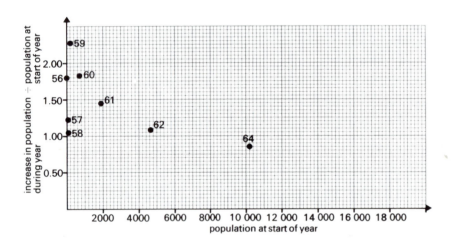

Figure 9   The annual growth of the collared dove population as a fraction of the population at the beginning of each year (from Table 6), plotted against the population at the beginning of each year.

Do you have a great deal of confidence in your estimate in SAQ 14?

When I set up the logistic equation (11), I commented that the supposition that the fractional growth rate declined *linearly* was quite arbitrary, so you should not have a great deal of confidence in your prediction. The estimate depends crucially on this supposition, but there is nothing in the data that particularly supports it. On the contrary, the data show a non-linear decline. Drawing a smooth curve through the data points and seeing where it meets the P-axis could produce a very different answer (Figure 10).

Though, in principle, I can estimate the maximum population of collared doves from Figure 10 in this way, it is inconvenient because the point I am looking for is well off my scale. I can overcome this problem by plotting the logarithms of the population along the horizontal axis instead of P. I have done this in Figure 11. I have used logarithms to the base ten simply because

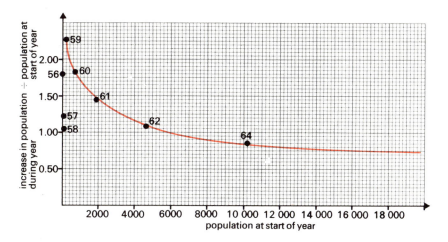

Figure 10   An attempt to estimate the equilibrium population of collared doves by drawing a smooth curve through the later data points of Figure 9.

they are more convenient than natural logarithms here. Interestingly, the data points now *do* lie roughly on a straight line, so I shall use a straight line as my model, as shown in Figure 11. (Of course, the *axis* is non-linear.)

SAQ 15

SAQ 15

Use Figure 11 to estimate the maximum population of collared doves in Great Britain.

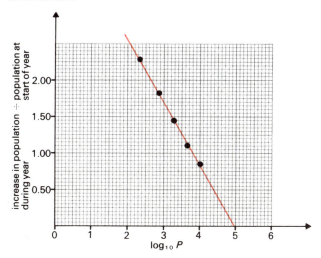

Figure 11   An estimation of the equilibrium population of collared doves. The logarithm (to the base ten) of the population is plotted horizontally, to compress the scale. It is then found that a straight line is a reasonable fit to the data points.

As the solution to SAQ 15 shows, this model gives a predicted equilibrium population of about 100 000. This prediction seems quite reasonable. Current estimates of the British population of collared doves are of this order of magnitude.

This model seems quite an accurate description of the growth of the collared dove population. Let us see what modification of the logistic equation (11) it corresponds to. I have estimated the equilibrium population on the supposition that the instantaneous fractional growth rate declines as a linear function of $\log_{10} P$ (see Figure 11). This corresponds to taking a function $f(P)$ on the right of equation (10) of the form

$$f(P) = a - b\log_{10} P \qquad \text{or} \qquad f(P) = a - c\ln P$$

where $\log_{10} P = \ln P/\ln 10$ and a, b and c are constants

So the differential equation

$$\frac{1}{P}\frac{dP}{dt} = a - c\ln P \tag{13}$$

seems a fairly accurate description of the later growth pattern of the collared dove population.

SAQ 16 (Unit 12)

Use the substitution $Z = \ln P$ to solve equation (13).

SAQ 17

Suppose you are using equation (13) as a model of population growth up to a maximum population level and that this maximum population is M. Express M in terms of a and c.

SAQ 18 (Unit 7)

Show that the population growth curve predicted by equation (13) has a point of maximum slope. Find this maximum value of the slope and the population level at which it occurs.

I have found, in equation (13), a description of the growth of the collared dove population that is more accurate than the logistic equation (11). If this, or any other particular equation, were *always* a more accurate description of population growth than the logistic one, then I should discard the logistic equation from my theory of population growth and replace it with this more accurate equation. However, equation (13) is no more universally accurate than the logistic equation. It seems more accurate for collared doves, but for another species the logistic equation could be the more accurate. In different cases, population growth may be more accurately described by various choices of the function $f(P)$ in equation (10).

Whether it is worthwhile to look for an equation that is more accurate than the logistic equation in a model depends on your problem. The logistic equation is the simplest of the general form needed to describe 'growth with limits' and it is often difficult to achieve accuracy in descriptions of population growth, with the data so affected by the random variations discussed earlier. You must remember that the logistic equation will probably not be suitable if your model is to be accurate. To predict the likely equilibrium population of collared doves I needed a more accurate model. An example where a logistic equation may be suitable is provided by the model used for maximizing the yield from a fishery, which was discussed in Unit 12, Section 2.2. The most important fact in this case is that the logistic curve has a point of maximum slope, the population level at which it occurs is less important. Variations of the logistic equation also have this property (SAQ 18 provides an example.)

The factors that actually come into play to inhibit the growth of a particular population as it reaches its limiting level are interesting. The most obvious possibility is that of starvation through food shortage. This is believed to be most important factor for some species: many species of birds are probably limited in this way by deaths during winter. There are, however, many other possibilities and in some cases they are certainly more important. Predation and disease have both been shown to be the most important factor for some species. The territorial behaviour of certain birds means that available space becomes a limiting factor. In many species of predatory birds (such as the tawny owl) the essential limitation is food, but this operates in an interesting way. If individual birds find food hard to find (either because the population of owls is high or because that of their rodent prey is low) then they are very much less likely to breed. As the tawny owl population approaches its limiting level, it achieves equilibrium by decreasing the birth rate, rather than simply keeping a constant birth rate and allowing the death rate to increase (as is the case if limitation is by starvation when food is short).

26

4   OSCILLATING POPULATIONS: HARE AND LYNX

So far I have considered two general types of population dynamics. First, I looked at a population in balance with its environment, whose numbers are roughly constant, but with an element of random fluctuation. I then considered a population increasing in numbers, with this growth levelling off as a maximum (or 'equilibrium') population was reached. However, some populations vary in a way that is different from either of these.

Figure 12 shows the variations in the population of a medium sized predator, the lynx, which lives in northern Canada. The striking feature here is the regularity of the variations; rising and falling in a ten-year cycle. The amplitude of the oscillations does vary, but the peaks are all ten years apart, as are the troughs. The extent of the variation of this population is more readily seen from a graph of the logarithm of the population (Figure 13). The population varies between each trough and a subsequent peak by a factor of between thirty and a hundred. (Compare this with the random variations of Section 2, where, for example, the pied flycatcher population shown in Figure 1 varies between its greatest and smallest levels by a factor of about two.)

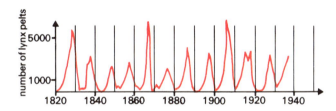

Figure 12   The number of lynx pelts sold by the Hudson's Bay Company from the Mackenzie River District, trapped in the years 1820–1934 (from Elton and Nicholson, 1942).

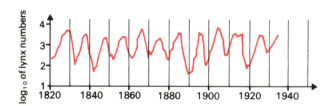

Figure 13   A graph of the logarithm of the lynx population of Figure 12.

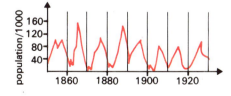

Figure 14   Population of the snowshoe hare. (Kormondy, E. J. (1959) Concepts of Ecology, Prentice-Hall p. 96. Redrawn from MacLulich, D. A. (1937), University of Toronto Studies, Biological Series No. 43. Based on records of pelts received by the Hudson Bay Company.)

Populations of some other species show similar cyclical variations. Figure 14 shows the population of a species of hare (the snowshoe hare) which is resident in the same area as the lynx. Figure 15 shows a possible British example, the ptarmigan, a grouse that lives on the high tops of the Scottish hills. The population of the Scottish mountain hare has a similar variation. Other examples may be found in the tundra of northern Europe—the lemming and various of its predators, for example, the long-tailed skua (Figure 20).

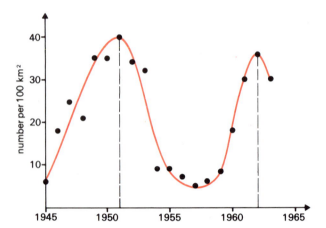

Figure 15   The population of the ptarmigan in the Cairngorms. The variations suggest a cycle of eleven or twelve years, although the variations in numbers here are much less dramatic than those of the hare and the lynx.

Although I have quoted a fair number of examples, I should not want to imply that this sort of oscillation in population size is common. It is in fact a tendency shown by very few populations. It seems likely that there is something special and unusual about these species; in their evolution or their relationships with their environments. My aim in this section is to try to use mathematical models to investigate the reasons why these populations show this striking oscillations.

Two points show up in the examples I have quoted. One is the sort of area from which the examples are drawn. They are remote areas, with relative simple ecosystems. Only a few species of animal and bird survive in northern Canada, in the tundra of northern Europe, or on the mountain tops of Scotland. Another feature is the relationship between *predator* and *prey*. Cyclical variations are evident in lemmings and in their predators. The lynx is a predator of the snowshoe hare; this species is in fact by far the most important food source for the lynx.

I shall start my investigation of the reasons for population oscillations by looking at a simple model of the interaction of a predatory species and its prey. I shall not be looking for *accuracy* in this, or other, models in this section. Rather, I shall be looking for the predictions of the model to correspond with the most general features of the data—their oscillatory nature; and I shall be investigating how variation of my modelling suppositions improves (or otherwise) this general correspondence of prediction with data.

4.1   A model of predator–prey interaction

In this section I want to set up a model of the interaction between a predator and its prey. I shall call the predator population L (for lynxes) and the prey population H (for hares), although I do *not* want to imply that the model is only intended for these two particular species.

As I am looking at the *interaction* between predator and prey as the possible cause of oscillation, I shall make modelling suppositions accordingly. I shall suppose that the prey species is the *only* food source for the predator. (This is certainly a reasonable supposition for hare and lynx.) Variations in the prey population then must have an affect on the predator population. I shall also suppose that the predator population affects the prey population.

In setting up a model, I will consider the factors affecting the hare population in two parts. I will suppose that, if there were no lynx predation, the hare population would grow logistically. This combines all the influences on the hare population except for predation, which I shall look at separately. In this model, I shall suppose that the lynx population depends *only* on the size of the hare population.

I shall suppose that a hare population of H_0 would be in equilibrium with a lynx population of L_0. At these population levels, I suppose, the lynxes eat just enough hares to stop the hare population increasing and that this provides just enough food to keep the L_0 lynxes alive, with nothing spare to allow the lynx population to increase. The next self-assessment question indicates that it may be reasonable to suppose that there *is* just one level at which the two populations are in balance.

SAQ 19

SAQ 19 (Difficult)

Suppose that, in equilibrium, the number of lynxes is proportional to the number of hares and that the number of hares eaten is proportional to the number of lynxes.

Also suppose that the hares would increase logistically but for predation.

Show that, on these suppositions, there is just one population level at which hare and lynx can be in equilibrium. (Hint: in equilibrium, the *rate of increase* of the hare population is zero.)

My next supposition is simply a choice of a special case in order to simplify the mathematics. I shall suppose that the equilibrium level H_0 of the hare population is somewhere near the middle of the logistic curve. I do this because it is not unreasonable (see Figure 16) to approximate the middle part of the logistic curve by a straight line. I want to see what happens if the hare and lynx populations are initially out of equilibrium, though not by so much that the hare population is outside that central, roughly linear, part of the logistic curve.

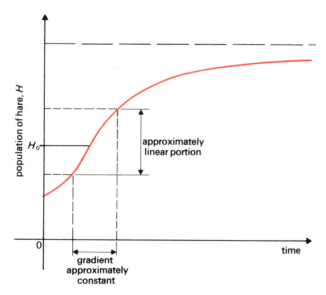

Figure 16   The logistic curve, modelling the increase of the hare population if there were no predation.

I can now write down some equations. Without predation, I suppose that the hare population would increase at a rate k, the slope of the middle of the logistic curve (which I am supposing constant). I will suppose that the number of hares eaten is proportional to the lynx population. So

$$\frac{dH}{dt} = k - aL$$

where the parameter a is *positive*, of course.

29

I know that the hare and lynx populations would be in equilibrium if L were L_0 and H were H_0. What can I learn from this?

If the hare population is in equilibrium, $dH/dt = 0$. Therefore $k = aL$ at equilibrium. However, $L = L_0$ at equilibrium, so the constant k equals aL_0.

Substituting this value for k gives

$$\frac{dH}{dt} = aL_0 - aL = a(L_0 - L) \tag{14}$$

This suggests to me that I introduce a new variable

$$l = L - L_0$$

This measures the extent to which the lynx population is above its equilibrium level. I shall also write

$$h = H - H_0 \tag{15}$$

so that h is the extent to which the hare population differs from its equilibrium size.

I want now to express equation (14) in terms of h and l. Substituting for l in equation (14) gives

$$\frac{dH}{dt} = -al$$

Differentiating equation (15)

$$\frac{dh}{dt} = \frac{dH}{dt}$$

Therefore

$$\frac{dh}{dt} = -al \tag{16}$$

This equation is the first part of my mathematical model.

Consider the lynx population. I supposed that this responds only to the number of hares. If there are many hares, the lynxes increase; if there are few hares, they decrease. A crude model of this could be

$$\frac{dl}{dt} = bh \tag{17}$$

where the parameter b is positive.

Equations (16) and (17) taken together are my mathematical model of predator–prey interactions. To solve the model, I would like to find an equation involving just one variable. I can do this. First differentiate equation (17) with respect to t. This gives

$$\frac{d^2l}{dt^2} = b\frac{dh}{dt}$$

From equation (16), $dh/dt = -al$, so

$$\frac{d^2l}{dt^2} = b(-al) = -abl \tag{18}$$

Remember that I noted when setting up the model that a and b are both positive.

SAQ 20 (Unit 9)

What can you say about the solution to equation (18)?

Solve equation (18) for l. Find the corresponding solution for h by substituting into equation (17).

The model predicts that the hare and lynx populations will oscillate. They *could* be in a steady equilibrium at $H = H_0$, $L = L_0$, but if they are disturbed from this equilibrium level by some random fluctuation, then they will not return to that level, but will oscillate about it. Of course, this is a very crude model and you may well feel that I made so many doubtful suppositions in setting it up that it is of no real significance. I do not claim that the suppositions I made should be expected to be borne out in reality just becaue I do have a prediction of population oscillations. Quite on the contrary; this could not be expected to be an accurate model of any real population. The point is that it lends support to the theory that it is *in the interaction of predator and prey* that the explanation of population oscillations lies. Remember, in setting up the model I supposed that there was interaction between the two species; that prey affects predator and predator affects prey. On these suppositions, I find that my model predicts oscillations, so this could be the explanation for the oscillation in the populations of hare and lynx. Of course, I have not proved that is: it is quite possible that wholly different suppositions would also lead to a prediction of oscillations.

I quoted earlier some examples in which populations of prey species and their predators both oscillate. It was these examples that prompted me to set up my predator–prey model. A more instructive example is that of the control in Australia of the prickly pear cactus. After being introduced, this species had increased in numbers so much as to become a pest. In an attempt to control it, a moth that feeds on the cactus was also introduced to the country. The result was, first, a dramatic reduction in the population of the cactus, followed by oscillations in the level of its population and that of the moth.

Although I set up my predator–prey model as a possible model of hare and lynx populations, there is no reason why it should not be applied to the interaction of other predatory species and their prey. In particular, it could be considered as a model of the interaction between the prickly pear cactus and the moth. It seems likely to be a reasonable representation of the interaction in this example, in a general way at least: the population of the cactus was not showing oscillations prior to the introduction of the moth, but subsequently did. Also, I can see that one of the suppositions that went into my model is reasonable in this case: that predator affects prey. This is not always the case; many predators do not take enough of the prey to affect significantly their population. The introduction of the moth certainly affected the cactus. The first thing that occurred—before any oscillation— was a considerable reduction in the population of the cactus.

My model suggests that oscillations are likely to arise in a situation where there is *interaction*. If each of two species—predator and prey–affects the population level of the other, then their populations are likely to oscillate. In the example I just discussed, the oscillations do seem to occur because of the interaction between the cactus and the moth. But do hare and lynx populations oscillate because of the interaction between the species? It is not *necessarily* the case that there is interaction here. The lynx *is* heavily dependent on the hare population—it is its main food source at all times, so the hare population certainly affects the lynx population. However, the lynx may not affect the hare—they may not eat enough to affect the hare population significantly.

There is one further way I can compare the predictions of my model with reality; by looking at the *phase* of the oscillations of the lynx and hare populations.

Two sinusoidal oscillations such as those in Figures 17 (a) and (b) are said to be *in phase*, because the peaks and their troughs always occur at the same time. An oscillation such as that in Figure 17(c) is $\pi/2$ *out of phase* with the first two. This is because the curve in Figure 17(c) has the equation

$$y = A \sin\left(\omega t + \frac{\pi}{2}\right)$$

while both of the curves in Figures 17(a) and (b) have equations of the form

$$y = B \sin \omega t$$

differing only in their *amplitude, B.*

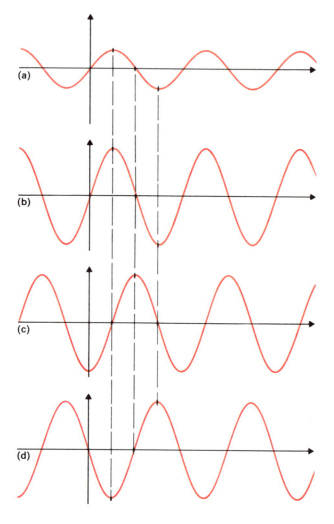

(a)

(b)

(c)

(d)

Figure 17   The idea of phase. The oscillations in (a) and (b) are in phase. The oscillation in (c) differs in phase by $\pi/2$ from those in (a) and (b). That in (d) is out of phase with those in (a) and (b) by π.
These oscillations all have the same frequency. The size of the amplitude of the oscillations is not important here.

The oscillation in Figure 17(d) is out of phase with those in (a) and (b) by π, because that has an equation of the form

$$y = C \sin(\omega t + \pi)$$

(The *amplitudes A, B* and *C* must always be positive.) Only oscillations with the *same* frequency, ω, can be compared for phase.

In the solution to SAQ 20 I arrived at the following equations for the hare and lynx populations.

$$\left. \begin{array}{l} h = \omega \dfrac{A}{b} \cos(\omega t + \alpha) \\[2mm] l = A \sin(\omega t + \alpha) \end{array} \right\} \tag{19}$$

SAQ 21

(a) Use the formula for sin $(\alpha + \beta)$ on page 16 of your *Handbook* to expand $\sin(\omega t + \pi/2)$. By how much do $\cos \omega t$ and $\sin \omega t$ differ in phase?

(b) Does the model predict that the oscillations of the hare and lynx populations are in or out of phase? If they are out of phase, by how much?

(c) How does this prediction of the model compare with reality?

The solution of my model—equations (19)—predicts oscillations of the hare and lynx populations, but oscillations out of phase by $\pi/2$. The period of the oscillations is about ten years and this period corresponds to a phase angle of 2π. So the model predicts oscillations out of phase in time by a quarter of this period, 10/4 years. (Two or three years in reality.) There is no evidence of a systematic lag of two or three years in the oscillations of the lynx population behind that of the hare. So far as one can tell from Figures 12 and 14, the peaks and troughs all correspond closely, so this prediction of the model does *not* correspond well with reality.

The model I set up in this section was a very crude one. Should I look at the predictions of such a model in this much detail? If the model is worth anything at all I do think this prediction, as to the phase of the oscillation, *should* correspond with reality. It is on the same general level as the prediction that the populations will be oscillatory.

The model set up in this section is of a situation where there is *interaction* between predator and prey—variations in each population significantly affecting the population of the other species. In this situation, the model predicts that each population will oscillate about its average level. The oscillations of predator and prey are, however, predicted to be out of phase (by $\pi/2$). The second prediction does not correspond with reality for the hare and the lynx. Before looking for explanations of the oscillations of these populations other than in their interaction, I shall first look at a slightly more elaborate predator–prey model. Will it predict oscillations that are in phase?

4.2 'Natural control'

My first model of the interaction of a predator and its prey (which I called lynx and hare for convenience) was set up in equations (16) and (17) of the previous section. In a more elaborate model I might modify both of these equations, although I shall change only one here.

In setting up equation (17) I supposed that changes in the lynx population depend only on the number of hares.

What other factor might I wish to take into account here?

The lynx population itself.

Returning to the beginning of my 'theory' of population increase (Section 3.1), the more lynxes there are, the more there are to breed. This idea leads to the equation $dl/dt = cl$ and the exponential model of population growth. To take this into account it looks as if I should modify equation (17) to

$$\frac{dl}{dt} = bh + cl \qquad (20)$$

where b and c are positive.

Is this the right approach?

It may well be too simple. At the end of Section 3.3, I mentioned the behaviour of a particular predatory species, the tawny owl. Research on a population of these birds has shown that the number of births depends more on the available prey, than on the owl population itself.

33

What equation might model a predator that behaves in this way?

Equation (20), but with c *negative*.

The rate of increase, dl/dt, of the predator will equal births minus deaths. If I suppose that births are proportional to the prey population and deaths proportional to the predator population, then I arrive at equation (20), with c negative.

Study comment

You may be worried here by the fact that the variables l and h represent not the actual lynx and hare populations, but their differences from a hypothetical equilibrium level. I could reach equation (20) by first setting up an equation

$$\frac{dL}{dt} = bH - bH_0 + cL - cL_0$$

in terms of the actual populations and then changing variables by $l = L - L_0$, $h = H - H_0$ as I did in Section 4.1, but I wanted to avoid that manipulation. The result—equation (20)—is the same.

Equation (20), then, may represent two possibilities. The equation with c negative is the model for a predator which behaves like the tawny owl in that it varies its breeding in response to available prey. I shall refer to this situation as one in which the predator has 'natural control' (self-control if you like), and take the size of $-c$ as a measure of the extent of this 'natural control'. The equation with c positive seems to represent a more straightforward situation.

In setting up the previous model, I supposed that the hare population at equilibrium, H_0, fell in the middle part of the logistic curve and I approximated the logistic curve by a straight line. The resulting equation was

$$\frac{dh}{dt} = -al \tag{16}$$

I could set up a model without looking at this special case. The modification needed is to introduce a term on the right of equation (16) involving h. However, the mathematics in the special case is simpler and it illustrates what is needed, so I shall continue to confine my attention to this case.

The parameter a is positive, as is b in equation (20). The sizes of the parameters a and b represent the amount of interaction between the two species—the extent to which hares affect lynxes and lynxes affect hares.

Equations (16) and (20) are my new model of predator–prey interactions. I must now solve the mathematics to find the predictions of this new model. An equation involving only l can be obtained by the same method as I used in the previous section.

SAQ 22

SAQ 22

Eliminate h from equations (16) and (20) to obtain a differential equation involving only l.

As shown in the solution to SAQ 22, I can obtain an equation involving only l: this is

$$\frac{d^2l}{dt^2} - c\frac{dl}{dt} + abl = 0 \tag{21}$$

What types of solution may this equation have?

Three cases are mentioned in your *Handbook*. The solution may take the form

(a) $l = Ae^{r_1 t} + Be^{r_2 t}$

(b) $l = (At + B)e^{rt}$

(c) $l = e^{rt}(P \sin \omega t + Q \cos \omega t) = e^{rt}[A \sin(\omega t + \epsilon)]$

My main interest here is in the possibility of oscillations

Under what conditions does equation (21) have *oscillatory* solutions?

When $c^2 < 4ab$.

SAQ 23 (Units 14 and 15)

SAQ 23

What is the solution of equation (21) when $c^2 < 4ab$?

For certain values of the parameters ($c^2 < 4ab$) my new model does predict population oscillations and in this case the predicted population variation is given by the equation

$$p = \exp\left(\frac{c}{2}t\right) \; [C \sin(\omega t + \alpha)]$$

where $\omega^2 = \frac{1}{2}\sqrt{(4ab - c^2)}$. This prediction differs from that of my previous model in that the amplitude of the oscillation is predicted to change with time.

What happens to the amplitude of the oscillation?

It either increases with time ($c > 0$); or it decreases with time ($c < 0$). (If $c = 0$, it is constant.)

These two cases have rather different interpretations in reality. If $c > 0$, an increasing oscillation is predicted. This implies that the predator population will eventually become extinct, since the actual predator population $P = p + P_0$ will become zero as the amplitude of the oscillations increases (see Figure 18(a)).

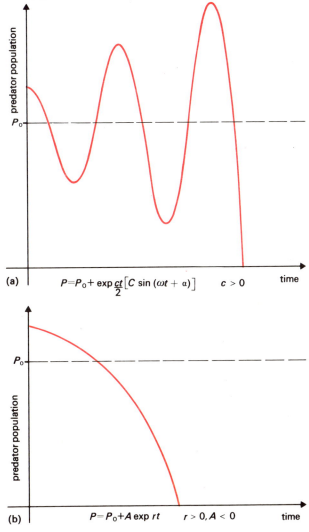

(a) $P = P_0 + \exp\frac{ct}{2}\left[C \sin(\omega t + \alpha)\right]$ $c > 0$ time

(b) $P = P_0 + A \exp rt$ $r > 0, A < 0$ time

Figure 18   Roads to extinction. Possible predictions of the model for a predator lacking 'natural control'.

35

If $c < 0$, then damped oscillations are predicted. The extent of damping depends on the size of $c/2$. Such damping will tend to take the population back to its equilibrium level and eliminate oscillations. However, a real population will be subject to frequent random disturbances taking it away from equilibrium. I would expect a population to show persistent oscillations in this case so long as the damping is small.

I have not yet looked at all possible values of the parameters a, b and c.

> What is the solution of equation (21) when $c^2 > 4ab$? How would you interpret the predictions of the model in this case?

In this case, the solution is a sum of exponentials

$$p = A \exp r_1 t + B \exp r_2 t$$

where $r_1 = \frac{1}{2}[c + \sqrt{(c^2 - 4ab)}]$ and $r_2 = \frac{1}{2}[c - \sqrt{(c^2 - 4ab)}]$. My interpretation of this prediction of the model depends on the signs of r_1 and r_2. If both of these are negative, the solution is a sum of negative exponentials and the model predicts a smooth non-oscillatory return of the population to its equilibrium level, if it is displaced from it. If either r_1 or r_2 is positive, however, the solution contains a positive exponential and this is the dominant term in the solution. If the solution contains a term of the form $A e^{rt}$, with $r > 0$, then the population is predicted either to decline to extinction (A negative, Figure 18(b)), or to increase exponentially (A positive).

> If the predator population increases exponentially, what happens to the prey?

It must decline to extinction. This can be seen formally by substituting the solution $l = A \exp rt$ into equation (16) to find the corresponding prey population (see SAQ 24). If the prey becomes extinct, the predator must do also. In this case, the prediction is that the predator population first increases exponentially, but then crashes to zero.

SAQ 24 (Unit 11) SAQ 24

If $l = A \exp rt$ and $dh/dt = -al$, what is h in terms of t?

Knowing that A, a, and r are all positive, sketch a graph of h against t.

The prediction I have just discussed is for the case when $c^2 > 4ab$ and either r_1 or r_2 is positive.

> When will either r_1 or r_2 be positive?

This occurs when c is positive. For if c is positive then $r_1 = \frac{1}{2}[c + \sqrt{(c^2 - 4ab)}]$ is positive. But if c is negative, then r_1 is in fact negative, and so is $r_2 = \frac{1}{2}[c - \sqrt{(c^2 - 4ab)}]$. ($r_1$ is negative in this case because r_1 is the sum of c—which is negative—and $\sqrt{(c^2 - 4ab)}$, which is positive, but must be smaller in magnitude than c.)

The predictions of the model for the predator population are summarized in Table 9.

I mentioned my interpretation of the parameters a, b and c when I set up the model. The case when c is negative corresponds to a predator with 'natural control'. The size of ab measures the extent of the interaction between the two species.

The model predicts that a predator lacking any natural control (c positive) will always become extinct. However, the temptation to draw conclusions from this prediction should be resisted. My conclusion is that the model of

Table 9

	$c^2 > 4ab$	$c^2 < 4ab$
c negative	returns to equilibrium smoothly without oscillation	damped oscillation damping depends on size of $c/2$.
c positive	solution contains a positive exponential predator becomes extinct	solution is an increasing oscillation predator becomes extinct

equations (16) and (20) cannot describe any *real* population in this case, for no real population is extinct. No real predator population could continue to increase its numbers even when its prey population was very low, but that possibility is a built-in supposition of equation (20) if c is positive.

In the case where the predator does have a degree of natural control, the predicted behaviour depends on the size of this, relative to the extent of interaction. Oscillations are most likely when the predator's natural control is small and the extent of the interaction large.

These predictions, of course, depend on the original suppositions made in setting up the model. Although the model considered here is a little more complicated than that in Section 4.1, it is still crude and inaccurate. For example, although it predicts population oscillations under certain circumstances, it predicts *sinusoidal* oscillations. The oscillations of hare and lynx shown in Figures 12 and 14 are not exactly sinusoidal! A model like this can lend some support to a theory already formed on other evidence, although taken on their own I cannot place very much credence in its predictions.

My purpose in setting up this model was to see if it predicted oscillations whose phases corresponded with those of the oscillations of the actual hare and lynx populations. Does this model do any better on this point than that in Section 4.1?

SAQ 25 (Units 9 and 15)

SAQ 25

In the case where the model predicts oscillations, the predicted lynx population is

$$l = C \exp(ct/2) \sin(\omega t + \alpha)$$

where $\omega = \frac{1}{2}\sqrt{(4ab - c^2)}$. Use the equation (20)

$$\frac{\mathrm{d}l}{\mathrm{d}t} = bh + cl$$

to find the predicted hare population, h.

Are h and l in phase?

The solution to SAQ 25 shows that the model predicts a hare population given by

$$h = \frac{C}{b} \exp\left(\frac{c}{2}t\right)\left[-\frac{c}{2}\sin(\omega t + \alpha) + \omega\cos(\omega t + \alpha) \right]$$

These oscillations are again out of phase with those of the lynx population. They are not so much as $\pi/2$ out of phase this time, because there is now a positive multiple of $\sin(\omega t + \alpha)$ in the solution for h. However, I noted that persistent oscillations are most likely when c is small (low damping) and ab large (making $\omega = \sqrt{(4ab - c^2)}$ relatively large). In this case the cosine term in this solution dominates and the predicted phase difference is nearer to $\pi/2$. Again the model predicts that the oscillations of hare and lynx should be two years out of phase, so this model fails to match reality on this point, just as did that in Section 4.1.

Is this a sufficient reason to reject these models completely?

No—I have an alternative; to reject the idea that the oscillations of the hare and lynx populations are caused by the interaction of these populations. Perhaps there is an alternative explanation, consistent with the predictions of these models. I shall investigate this question next.

4.3 Conclusion

In Sections 4.1 and 4.2, I described models of the interaction of predator and prey that predict population oscillations, at least under certain circumstances; but they predict oscillations that are out of phase, which does not correspond to the data for the hare and the lynx.

> If the interaction of these two species is not the cause of their oscillations, what other explanation might there be?

It might be that the basic cause of the oscillation is in the interaction between the hare and *its* food source. The predator–prey models I have set up can just as well be used to describe this interaction. If this is the cause, then it is the oscillation of hare and its food supply that should be out of phase.

> How might this hypothesis be tested?

Investigate the population of the hare's food source. This is vegetable, but plant populations can be measured. If the hypothesis is correct, the hare should be dependent on one (or perhaps a very few) species for its food; and the population of this food species should vary cyclically, with the cycles *not* in phase with those of the hare.

> If this is correct, why does the lynx population oscillate?

Because the hare population oscillates, and the lynx depends on the hares. One might then regard the lynx as being in a state of 'forced oscillation', but the necessary mathematics has not been considered in this course. However, such a model again predicts oscillations of the lynx population and again these oscillations are out of phase with those of the hares. This model could only describe a real population with c negative, for with c positive it predicts extinction (as seen in Section 4.2). The model with c negative is set up on the supposition that the predator has 'natural control': that it varies its birth rate in response to variations in the prey population. Perhaps the lynx does not behave in this way. Perhaps, rather, it breeds as fast as it can when food is plentiful and many lynxes starve when the hare population falls. If the lynx can breed so fast that there are always the number of lynxes that can survive on the existing hares, then

$$L \propto H$$

is the appropriate model. If the hare population oscillates, then this model certainly predicts that the lynx population will oscillate. It fits the data so far as phase goes too!

A pair of lynxes typically have a litter of two cubs, so the population has the potential to double each year. If the lynxes do just breed flat out when there is plenty of food and starve when food is short, then they might be expected to increase exponentially during 'good times': perhaps doubling each year. If this *is* the case and I draw a graph of the logarithms of the lynx population, it should look like Figure 19. In the five years of increase, the population would increase by a factor of $2^5 = 32$, then fall again, and

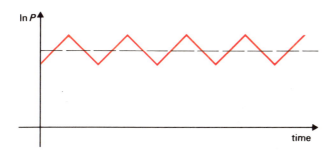

Figure 19 A saw-tooth model of population oscillations. (Note that the logarithm of the population is being plotted.) The population increases and decreases at a constant rate.

so on. A plot of the logarithms of the actual lynx population was shown in Figure 13. This bears quite a close resemblance to Figure 19, and the lynx population *does* rise by a factor of the magnitude of thirty during each period of increase. During some periods of increase (such as that from 1890) it increases by clearly *more* than this. A possible explanation of this is that the oscillations are asymetric, taking six years to rise and only four years to fall. Then an increase by a factor of sixty-four is possible.

It seems likely that the lynx population does simply follow the hare population. The underlying reason for the oscillation of these populations must lie elsewhere—not in the interaction of the hare and the lynx. An interaction of the predator–prey type between the hare and its food supply seems most likely. In Section 4.1, I concluded that oscillations can be expected where there is interaction. That the interaction of the hare and the lynx does *not* cause oscillations is only consistent with this idea if that interaction is only one way. So I would also suggest that the lynxes do not take enough hares to affect significantly the hare population.

Models such as those in Sections 4.1 and 4.2, that predict population oscillations, are usually referred to as predator–prey models. However, the examples I have considered do not suggest that oscillations in the size of populations are usually caused by the interaction with their prey of those species that we would normally think of as predators: that is, species that prey on other animals or birds. Some such species have evolved 'natural control' that tends to damp out oscillations (though this does not seem to be the case for the lynx.) However, all species feed on something and so may in a sense be regarded as predators with a prey of their own. The interaction between a species and its food supply does seem to be the cause of population oscillations in some examples; but in the more convincing examples, the 'prey' is vegetable rather than animal.

The interaction of the hare with its food supply seemed a more likely explanation of the oscillations in that case than the interaction of the hare and the lynx. The interaction described between the moth and the prickly pear cactus is one example where the cause does seem clear. Another interesting example is the behaviour of the populations of several American deer species (elk, moose, white-tailed deer and north American mule deer) *after* their predators had been exterminated (by man). The deer species increased dramatically—previously their numbers had been controlled by the predators. The deer began to over-use and destroy their habitat and food supply, and their numbers fell again through starvation. This situation seems potentially oscillatory; to avoid an oscillation setting in, man may need to take over the role previously filled by the predators. A possibility is that these deer species have evolved in a situation where they were controlled by their predators and so had not evolved a 'natural control' which would eliminate the danger of oscillation.

If the interaction of *any* species with its food supply is a potential source of population oscillations, then these might be expected to be rather common, but this is not the case; they are quite unusual. I have mentioned that predators may not significantly affect the population of their prey species. A bird feeding on acorns is unlikely to affect the population of oak trees; nor are blackbirds feeding on worms likely to catch enough to affect the worm population. The 'natural control' behaviour of some predatory species also helps to minimize the extent to which they affect the population of their prey.

In an ecosystem with a wide variety of species, such as that of Britain, most species have many alternative food sources. In this situation interaction is less likely. Prey is less likely to affect predator, as it can turn to an alternative food supply if one prey species becomes scarce; and predator is less likely to affect prey, because it will concentrate on whichever food species is commonest, rather than overeating a single food source. Certainly the predator–prey models I described in this section will be inappropriate in a situation of many interdependent species, for they describe a simple situation of two species each mainly dependent on the other.

The model is only likely to be appropriate in a simple ecosystem, with few species. Interestingly, as I noted at the beginning of the section, many of the examples of population oscillations do come from such simple ecosystems. Complexity does seem to eliminate the likelihood of oscillation.

Even in a complex ecosystem, a predator at the top of the food chain may still be dependent on one, or just a few, food species. The tawny owl, for example, may prey almost entirely on two or three species of rodents. Such species might seem likely to exhibit population oscillations. That they do not is probably due to the 'natural control' behaviour they have evolved.

SAQ 26

SAQ 26

The logistic equation will never predict oscillations in population size. Can you identify a supposition that went into setting up that model that overlooks the possibility of oscillations?

SAQ 27

SAQ 27

Figure 20 gives information on the populations of a predatory bird, the long-tailed skua, and its rodent prey. The skua is a maritime bird, so adults who do not breed in times of food shortage at their breeding grounds are in no danger of starvation; they will return to sea to feed, but fail to breed.

(a) Suggest a mathematical model to predict variations in the total population of the long-tailed skua, indicating the suppositions that go into setting it up.

(b) Would you expect the skua population to show oscillations? (You are not expected to be able to solve the mathematical model you set up in (a). Just put down your best guess as to the nature of the solution.)

(c) Would you say that this bird shows 'natural control'?

I have covered by no means all of the ideas in population modelling in this unit that would have been relevant and mathematically accessible. For example, I have discussed models of the interaction between predator and prey; but there are many other forms of interaction between the populations of two species that I could have considered. The interaction of a

40

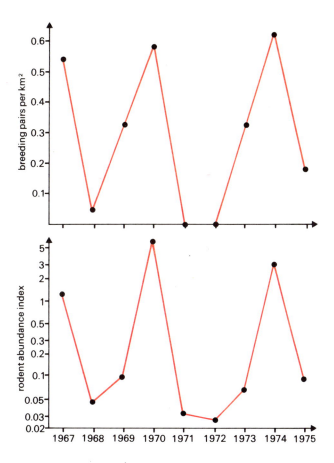

Figure 20 The number of pairs of the long-tailed skua breeding at a site in northern Sweden and the abundance of prey, between 1967 and 1975. Many adult birds fail to breed when prey is scarce, so the number of breeding pairs is not an accurate reflection of the total population in this case.

parasite with its host could be modelled by the same equations are a predator–prey interaction, the parasite having an adverse effect on the host. What of the types of association that are mutually beneficial to both organisms, such as the ox pecker and the animals with which it lives? (This type of interaction is called *symbiosis*.) And what of competition—the relationship between grey and red squirrels? Can the hypothesis that competitive species cannot coexist be supported by mathematical models? These questions could be examined using models like those considered in the text—but I had to leave *something* out!

In some ways the predator—prey models in this section are wildly inaccurate. The oscillations of the hare and the lynx (Figures 12 and 14) are very far from being sinusoidal. Also, it is not really reasonable to use the same equation to describe the lynx population during its periods of increase and those of decrease. Such crude models can give some insight into the relationships between populations; but they need to be used with care.

5 POSTSCRIPT

Finally, to conclude this course, I would like to take you right back to some of the modelling ideas expressed in the first unit. This final unit has attempted not only to revise some of the mathematics that has been taught in the course, but also to show again how this mathematics can help to describe and understand the behaviour of things in the real world. In short, it has been revising both the mathematical and the modelling aspects of the course.

As far as the mathematics is concerned, the course has been primarily concerned with teaching you how to deal with calculus: that is, differentiation, integration and the solution of certain differential equations. In order to reach this goal, various aspects of algebra, geometry, and so on, have also been covered. The intention has been to introduce you to this most important mathematical tool and to give you practice in its use in a number of practical situations. However, calculus is only one of many powerful mathematical tools available as you will discover if you choose to take further courses containing mathematics. Few aspects of mathematics, however, have such a widespread use as calculus, so I make no apology for concentrating so much time and effort upon it.

The key ideas about modelling, however, will not change much as you take more advanced courses.

I think there are two main points to hold firmly in mind.

Simplification for a purpose

In order to produce mathematical models the world has always to be simplified in some way or other. The simplification chosen is to some extent a matter of judgement and the purposes may be quite different on different occasions. The main constraint is often that of producing a mathematical equation which can be solved without too much difficulty— like, for example, the use of the logistic equation for collared doves (equation (11)). In this case, there can be little doubt that the function $f(P) = a(1 - P/M)$ does not *exactly* represent the growth rate of collared doves; indeed, collared doves are discrete entities, so that a continuous model (which suggests that doves might come in a steady flow, like water) is itself a simplification. However, even if we accept a continuous model to describe a discrete situation, there remains the further simplification of supposing that the decrease in the population growth rate is linear. The purpose of both simplifications is to obtain a mathematical function which is simple and can be fitted with reasonable accuracy to the data we have on collared doves, and which thereby enables us to calculate dove populations at other times or to predict future trends.

With collared doves, or with any population problems, the quantity to be modelled is usually clear; it is the number of individuals of the selected species. But if you measure a population this way you ignore the variation between individuals. In the fisheries models (TV 8) the difference between catching a big fish and a small fish may be too important to ignore, so the population is often measured by its mass. In other fields the quantity to be modelled is not always apparent and part of good modelling consists in choosing which important quantity to represent mathematically in order to obtain the information you want. For example, in economics, although the flow of money is, in the end, what matters, it may be better to produce mathematical models which refer to concepts such as supply and demand, or productivity. Equally, when the aim is to calculate an object's move-ments or displacements, the quantities to model may well be forces such as air resistance, the thrust of engines, and so on. Of course, although you may know very well that these forces are not constant or, for example, exactly proportional to velocity, it may make for a successful mathematical model to suppose that they are.

So the first general point about modelling is that simplification for a purpose is inevitable and indeed is the essence of the whole process. Associated with this point is a much less tangible one; good and bad models can only be so judged in relation to the purposes for which they were created. An accurate model may not necessarily be a good one. If its purpose is to arrive quickly at approximate predictions, a simpler or more manageable one may be better. On the other hand, there is no doubt that in many situations accuracy is essential and if the way to achieve accuracy is to set up a complicated model carefully reflecting the details of reality, then this is the price one must pay. On the whole, the Course Team has avoided setting up models of this type in the course; not because they are unusual or unimportant, but because setting up such models requires a detailed knowledge of the area under discussion and also because there was just not room in the course.

A simple example of a situation where accuracy is necessary could be a health authority planning future requirements for hospital beds. Simple extrapolation of past requirements would provide some sort of estimate here, but before making a substantial investment in a new hospital or closing an existing one, a more detailed analysis—providing a prediction of greater accuracy—would be advisable. For example, examination should be made of trends in the *structure* of the local population (the proportion of women of child-bearing age, the proportion of old people); of trends in the *health* of the population as a whole; of plans for the *development* of the region. Such details will complicate the calculation required for making a prediction, but their inclusion will provide valuable accuracy. The actual mathematics here would remain quite simple, but in some cases the goal of accuracy can make the mathematics very difficult. In yet other cases, mathematical flexibility is to be preferred even if some accuracy is sacrificed. Which is needed is a matter of judgement.

Empirical and theoretical models

When establishing mathematical models it is always worth trying to keep the distinction between empirical models and theoretical models clearly in mind. When both kinds of model result in the same mathematical expression, one's confidence in the usefulness or validity of both models is greatly increased.

An empirical model is one in which you draw a curve through the data you have measured and plotted on graph paper. The linear models I used to predict the human population of Great Britain in answering SAQ 2 (see Figures 21(a) and (b)) are examples of empirical models. I might have decided that a smooth curve like that of Figure 21(c) was a better fit to the data and used that as my model. This is again an empirical model. If I recognize this curve as being exponential in shape, I can call on the *theory* of population growth that I developed in Section 3.1. Choosing the para-meters A and r in the equation

$$P = A \exp rt$$

so as to fit the data, is more than an empirical model—it has some theoretical backing.

In Section 3.3, I found that the growth rate of the doves could be fairly accurately represented by the equation

$$f(P) = a - c \ln P$$

Although this provides a good fit to the empirical data, I can find no theoretical reason why it might be true; with the consequence that I am unable to suggest that this might be a common pattern in population trends. Thus, for collared doves, I only had an empirical model, unsupported by a theoretical one. So the second key point to bear in mind is that there are two alternative approaches to the task and that a combination of the two is usually worthwhile whenever possible. You should not only try to ensure that your mathematical model is a good enough fit to the relevant data, but also to understand why the model you draw up takes the form it does.

These then are the main points to keep in mind whenever mathematical modelling for practical purposes is undertaken. In more advanced courses the range of mathematical techniques you use may increase, but the modelling strategy I have summarized above will remain unchanged. This is true whether the mathematical models are concerned with technological calculations affecting the design of systems or of electrical or mechanical equipment, or whether they are concerned with predictions and decisions about populations, prices or, for that matter, any process of change.

I hope that the experience and understanding in both modelling and mathematics we have given you in this course will stand you in good stead in any courses requiring these skills that you care to take in the future.

ANSWERS TO SELF-ASSESSMENT QUESTIONS

SAQ 1

You learnt in Unit 1 that a Gaussian curve falls away almost to zero at three standard deviations on either side of its mean. The model therefore predicts that the pied flycatcher population will always fall within three standard deviations of the mean. That is, I expect that the population, p, will be such that

$$74.3 - 3 \times 14.6 \leq p \leq 74.3 + 3 \times 14.6$$

or

$$30.9 \leq p \leq 118.1$$

I should not really say that the population always falls between these values, but rather that it does so 'virtually always' (I shall return to this point in the text). I can confidently predict that the population will be between 31 and 118.

SAQ 2

I can see from Figure 3(a) that I will be able to draw a straight line that will go close to all the data points, if not exactly through them all. I shall set up a model on this basis. I can only make certain that it will fit two of the data points exactly. I will choose a linear model that fits the first and last data points, that is, for 1821 and 1891. I will let T be the number of years since 1821, and $P \times 10^7$ be the population at time T years.

So I want

$$P = 1.4 \text{ at } T = 0$$

and

$$P = 3.3 \text{ at } T = 70$$

The linear equation that fits these conditions is

$$P = 1.4 + \frac{T}{70}(3.3 - 1.4)$$

$$= 1.4 + 0.027T \qquad \text{(to two significant figures)}$$

The graph corresponding to this equation is shown in Figure 21(a).

The year 1911 corresponds to $T = 90$, so this model predicts a population in 1911 of

$$(1.4 + 0.027 \times 90) \times 10^7 = 38 \text{ millions} \qquad \text{(to the nearest million)}$$

This is not the only reasonable model for solving this problem, so you are not necessarily wrong if you have a slightly different answer. There are other linear models that fit the data just as well as the one I have used and there are reasonable non-linear models as well (one is illustrated in Figure 21(c)).

I shall look at just one alternative approach to solving this problem. The population increase in the various ten year periods are as shown.

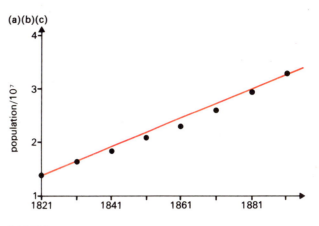

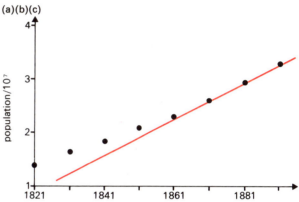

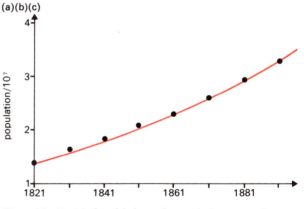

Figure 21 Empirical models for predicting the human population of Britain from the data in Table 2. (a) and (b) Linear models used in the solution of SAQ 2. (c) a non-linear curve fitted to the data points.

Table 10

period	1821–31	1831–41	1841–51	1851–61	1861–71	1871–81	1881–91
population increase in millions	2.5	2	2.5	2	3	3.5	3.5

These figures suggest that the rate of increase is gradually increasing. This time I will base my model on the supposition that the increase in each ten-year period in the future will equal 3.5 million, the increase in the last two periods. (This is equivalent to fitting a straight line through the last three data points—those for 1871, 1881 and 1891, as in Figure 21(b).) On this basis, my prediction for 1911 is a population, in millions, of

$$33 + 2 \times 3.5 = 40.$$

Either 38 million or 40 million, or something in between based on another straight line that fits the data reasonably well, is a satisfactory answer to the SAQ. Which is preferable depends on which underlying supposition you think more credible; whether you believe that the rate of increase in the future is more likely to equal that of recent years, or the average of the whole period for which data is available.

SAQ 3

(a) The figures 16.2, 17.3, 17.0 and 20.1 are measures of the 'error' involved in using as your model either one of the three lines l_1, l_2, l_3 or 'population is constant at 51.4'. The smallest error is 16.2, so l_1 is the best fit to the data by this criterion. The difference between 16.2 and 20.1 is reasonably significant, too, so I would think it worthwhile in this case to prefer the model predicting a linear increase to the simple model of random variation about the mean. I would therefore choose the model l_1 which is described by the equation

$$p = 30 + \tfrac{3}{2}t.$$

(b) In 1978, $t = 66$, so my prediction, using l_1, is

$$p = 30 + 99 = 129.$$

I am not very confident of this!

(c) I would prefer to predict a range of values within which I am reasonably sure that the population will lie. To make such a prediction, I will suppose that the 'errors' (the differences between the actual population value and those predicted by the linear model) have a Gaussian distribution with a standard deviation of 16.2 and a mean of 0. I can then be 95% confident that the population will fall within two standard deviations of the mean. I will then be 95% sure that the population will lie within the interval

$$129 - 2 \times 16.2 \leq p \leq 129 + 2 \times 16.2$$

That is, I predict that the population in 1978 will be between 96 and 162 pairs. (I have been cautious here, rather than round to the nearest whole number.)

SAQ 4

In 1963 the heron population suffered a substantial decrease below its normal level, because of a bad winter. This has happened on previous occasions—in 1942 and 1947. In both those instances the population subsequently increased, and I would expect this to have happened again in 1964.

SAQ 5

(a) The instantaneous fractional growth rate, r, is given by the equation

$$r = \frac{1}{p}\frac{dp}{dt}$$

Therefore r has the dimensions of $[\text{time}]^{-1}$.

(b) If the population, p, trebles each year, after t years

$$p = a \times 3^t$$

where a is the population at $t = 0$. If this population has an instantaneous fractional growth rate r per year then

$$p = ae^{rt}$$

Therefore

$$3^t = e^{rt},$$

Taking logarithms of each side and using the properties of logarithms

$$rt = \ln 3^t = t \ln 3$$

Thus the instantaneous fractional growth rate here is ln 3 per year.

Although I have not used capitals, I have been working in a particular set of units in part (b), and r and t represent dimensionless quantities.

Note, this is not the only way of arriving at this result; there are other correct ways of reaching this answer.

(c) If a population p has instantaneous fractional growth rate r, then

$$p = ae^{rt}$$

Suppose this population takes a time t_1 years to double. Since $p = a$ at $t = 0$, then $p = 2a$ at $t = t_1$. Therefore

$$2a = ae^{rt_1}$$

Thus $2 = e^{rt_1}$, and so $rt_1 = \ln 2$. The 'doubling time' is thus $\dfrac{1}{r}\ln 2$.

I can check the dimensions here. I know from part (a) that r has the dimensions of $[\text{time}]^{-1}$, so $\dfrac{1}{r}\ln 2$ has the dimensions of $[\text{time}]$, as it should.

SAQ 6

To examine whether an exponential model is likely to be suitable, I shall look at the increase in the world's human population in each ten year period as a percentage of the population at the beginning of each period. Using the data in Table 5, and working to two significant figures, I get

increase 1930–40 = 11%

increase 1940–50 = 8%

increase 1950–60 = 20%

increase 1960–70 = 22%

These figures are not constant, but the variation in them is no greater than the variation in those for the collared dove in Table 6. Again, I would think that an exponential model fits these data reasonably well—certainly better than a *linear* model.

Obviously, I expect to have a degree of inaccuracy in a prediction of the world's human population based on an exponential model fitted to these data, since no exponential will fit them exactly.

SAQ 7

(a) I want to fit an exponential model to the data in Table 4. To do this I have plotted the natural logarithms of those data as shown in Figure 22. I can draw a straight line that fits the data quite well and I have done so. Remember, plotting logarithms and drawing a straight line is equivalent to fitting an exponential model, because if

$$P = Ae^{rt}$$

then

$$Z = \ln P = \ln A + rt$$

There is more than one way of drawing a straight line through the data points in Figure 22. I have chosen one that fits slightly better the data for later dates than those for earlier dates, as I think they are more important in a model for prediction.

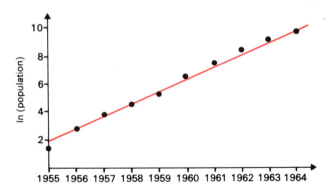

Figure 22 Prediction of the British collared dove population with an exponential model. To find the growth rate, a straight line is fitted to the logarithms of the population sizes.

The equation of the straight line I have drawn is

$$Z = 2 + \frac{8}{9}T$$

where Z is the logarithm of the population in the year $1955 + T$. My prediction for 1966 is given by $T = 11$ and so is

$$Z = 2 + \frac{8}{9} \times 11$$

$$= 2 + 9.778 \qquad \text{(to four significant figures)}$$

$$= 11.78 \qquad \text{(to four significant figures)}$$

The predicted population in 1966 is

$$P = \exp 11.78$$

Therefore, to the degree of accuracy to which I am working, my prediction of the 1966 population is 130 000

My prediction for 1977 is given when $T = 22$, so it is

$$Z = 2 + \frac{8}{9} \times 22$$

$$= 2 + 19.6$$

$$= 21.6.$$

Therefore

$$P = \exp 21.6$$

$$= 2.5 \times 10^9$$

I therefore predict a population of about 2 500 000 000 birds in 1977.

(b) The first prediction looks reasonable enough, but I do not believe the second one. If there were that many collared doves around, there would be about a hundred times as many of them as there are sparrows! I cannot use this model to make predictions that far ahead.

You may well have found quite different answers here, because fitting a slightly different straight line can give quite different results. There is nothing wrong if you read predicted answers for Z from your graph, rather than finding an equation for it.

SAQ 8

The instantaneous fractional growth rate of the human population, P, is approximately the difference between the birth and death rates. Using the suppositions in the text, this is

$$B - D = 0.04 - (0.04 - KT)$$

$$= KT \qquad \left(\text{for } 0 \le T \le \frac{0.04}{K}\right)$$

Now the instantaneous fractional growth rate is $\dfrac{1}{P}\dfrac{dP}{dT}$. So I get

$$\frac{1}{P}\frac{dP}{dT} = KT$$

This is the differential equation required.

SAQ 9

The equation

$$\frac{1}{P}\frac{dP}{dT} = KT$$

is solved by 'separation of the variables'. First rearrange the equation.

$$\frac{dP}{P} = KT dT$$

Taking integrals of each side

$$\int \frac{dP}{P} = \int KT dT$$

These integrals are 'standard' and are in the table of integrals in the *Handbook*.

$$\ln P = \tfrac{1}{2}KT^2 + c.$$

where c is the arbitrary constant. (Remember that only one is needed when solving differential equations by this method.) Taking exponentials of each side gives

$$P = \exp(\tfrac{1}{2}KT^2 + c)$$

$$= \exp \tfrac{1}{2}KT^2 \times \exp c$$

$$= C \exp \tfrac{1}{2}KT^2$$

where $C = \exp c$ is now the arbitrary constant.

SAQ 10

(a) 2010 is 160 years after 1850, so I want $U = 160$. Thus

$$\ln \frac{P}{P_0} = -8.6 \times 10^{-2} + \tfrac{1}{2} \times 1.1 \times 10^{-4} \times 200^2$$

$$= 2.2 - 0.086$$

$$= 2.1$$

Therefore

$$P = P_0 \exp 2.1$$

$$= 8.2 \times 1000 \text{ million}$$

$$= 8200 \text{ million}$$

(b) The rate of increase could be found from equation (9) by differentiation. This is not really necessary though. I can go back to the original differential equation, set up in SAQ 8. That gives the instantaneous fractional growth rate

$$\frac{1}{P}\frac{dP}{dT} = KT$$

I know K and in 1975 $U = 125$, and so $T = U + A = 165$. Hence the instantaneous fractional growth rate in 1975 was

$$1.1 \times 10^{-4} \times 165 \text{ per year} = 1.8 \times 10^{-2} \text{ per year}$$

to two significant figures: that is, 1.8% per year.

(c) If a population grows exponentially at an instantaneous fractional growth rate r, then it doubles in a time

$$\frac{1}{r} \ln 2$$

(Look at the solution to SAQ 4, part (c), if you are puzzled by this.) Hence the time for the population to double on the simple exponential model proposed is

$$\frac{\ln 2}{1.8 \times 10^{-2}} \text{ years} = 38 \text{ years}$$

Notice that this exponential model predicts a slightly longer doubling time than the 'exponential-plus' model.

(d) Remember that an upper limit was placed on the range of values of time for which the model is applicable in order to avoid supposing that the death rate becomes negative. This upper limit is

$$U = \frac{0.04}{K} - A = 364-40$$

$$= 320 \qquad \text{(to two figures)}.$$

That is, the instantaneous fractional growth rate will increase until 320 years after 1850, that is, until 2170.

SAQ 11

The logistic equation can be written

$$\frac{dP}{dt} = aP\left(1 - \frac{P}{M}\right)$$

I can solve it by inverting the equation and integrating with respect to P.

$$\frac{dt}{dP} = \frac{1}{aP(1 - P/M)}$$

47

and

$$t = \int \frac{dP}{aP(1 - P/M)}$$

This integral is not in the table in your *Handbook* so a substitution is needed. If I put

$$Q = \frac{M}{2} - P$$

then

$$P = \frac{M}{2} - Q$$

and

$$aP\left(1 - \frac{P}{M}\right) = a\left(\frac{M}{2} - Q\right)\left[1 - \left(\frac{1}{2} - \frac{Q}{M}\right)\right]$$

$$= a\left(\frac{M}{2} - Q\right)\left(\frac{1}{2} + \frac{Q}{M}\right)$$

$$= a\left(\frac{M}{4} - \frac{Q^2}{M}\right)$$

Also

$$dQ = -dP$$

Making the substitution in the integral gives

$$t = \int \frac{-dQ}{a\left(\frac{M}{4} - \frac{Q^2}{M}\right)}$$

With a little algebraic rearrangement, I can now use the table in your *Handbook*.

$$t = -\frac{M}{a}\int \frac{dQ}{\frac{M^2}{4} - Q^2}$$

$$= -\frac{M}{a}\int \frac{dQ}{\left(\frac{M}{2}\right)^2 - Q^2}$$

$$= -\frac{M}{a} \times \frac{1}{M}\ln\left(\frac{\frac{M}{2} + Q}{\frac{M}{2} - Q}\right) + c$$

where c is the arbitrary constant. Since $Q = M/2 - P$

$$t = -\frac{1}{a}\ln\left(\frac{M - P}{P}\right) + c$$

Finally, I will rearrange this equation to make P the subject

$$\ln\left(\frac{M - P}{P}\right) = ac - at$$

$$\frac{M - P}{P} = e^{ac} \times e^{-at} = Ce^{-at}$$

where C is a new arbitrary constant. Therefore

$$M - P = PCe^{-at}$$

$$M = P(1 + Ce^{-at})$$

Thus

$$P = \frac{M}{1 + Ce^{-at}}$$

This is the general solution to the logistic equation. (You could check it by differentiating and substituting back, if you like, but it is quite a messy calculation.)

I must choose C so as to fit the condition $P = P_0$ when $t = 0$. Substituting these values into the general solution gives

$$P_0 = \frac{M}{1 + C}$$

Therefore $C = \frac{M}{P_0} - 1$

This gives the solution

$$P = \frac{M}{1 + \left(\frac{M}{P_0} - 1\right)e^{-at}}$$

You may have found the solution in a rearranged form. There are various correct versions: for example

$$P = \frac{MP_0e^{-at}}{M + P_0 - P_0e^{-at}}.$$

SAQ 12

There is nothing in the model that confines it to the future, that is, to positive values of t. On the contrary, if I think it is reasonable to say that the population will *continue* to grow logistically, in order to estimate future populations, then surely it is equally reasonable to say that it *has been* growing logistically and use this supposition to estimate past populations.

SAQ 13

Equation (11) is

$$\frac{1}{P}\frac{dP}{dt} = a\left(1 - \frac{P}{M}\right)$$

The slope, S, of the logistic curve is

$$S = \frac{dP}{dt} = aP\left(1 - \frac{P}{M}\right)$$

To find a point of maximum slope, I will calculate dS/dP, and look for points where this is zero. Using the product rule

$$\frac{dS}{dP} = a\left(1 - \frac{P}{M}\right) - \frac{aP}{M}$$

$$= a - 2\frac{aP}{M}$$

So

$$\frac{dS}{dP} = 0 \text{ if } P = \frac{M}{2}.$$

I can see that this is a point of *maximum* slope, because the graph of S against P is a parabola opening *downwards*, since $S = aP(1 - P/M)$ is a quadratic equation with a negative coefficient of P^2.

SAQ 14

As you can see from Figure 23, you can draw graphs that suggest values between about 14 700 and 22 000. However, no answer below 18 855 is reasonable, since the population has already reached the level (Table 4). Even the larger of these estimates seems rather low when I look at Table 4 and see how fast the population was still increasing in 1963. I would choose the larger of these predictions, 22 000, as my estimate of the maximum population on this model.

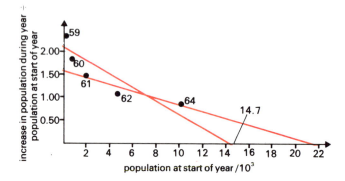

Figure 23 The maximum British population of collared doves, estimated by supposing that the fractional growth rate declines linearly with population increase, in accordance with the logistic model.

SAQ 15

The equilibrium population corresponds to zero growth and that occurs where the graph in Figure 11 meets the horizontal axis. That happens at when $\log_{10} P = 5$. The maximum population, M, is thus given by the equation

$$\log_{10} M = 5$$

Hence

$$M = 10^5 = 100\,000$$

I estimate that the British population of collared doves will settle at about 100 000.

SAQ 16

I shall make the substitution

$$Z = \ln P$$

directly into the differential equation

$$\frac{1}{P}\frac{dP}{dt} = a - c\ln P$$

rather than first rearranging the differential equation as an integral.

Differentiating with respect to t gives

$$\frac{dZ}{dt} = \frac{1}{P}\frac{dP}{dt}$$

Substituting in the differential equation gives

$$\frac{dZ}{dt} = a - cZ$$

To solve, I shall invert and integrate the equation.

$$\frac{dt}{dZ} = \frac{1}{a - cZ}$$

$$t = \int \frac{dZ}{a - cZ}$$

This is not quite a standard integral. It can be evaluated by the substitution $Y = a - cZ$, but perhaps you can do it in your head. If you cannot, refer back to Unit 12, Section 2.1. Integrating gives

$$t = -\frac{1}{c}\ln(a - cZ) + A$$

where A is the arbitrary constant. Rearranging

$$\ln(a - cZ) = Ac - ct$$

Therefore

$$a - cZ = \exp(Ac - ct)$$
$$= \exp Ac \times \exp(-ct)$$
$$= Be^{-ct}$$

where B is the new arbitrary constant. More algebra:

$$cZ = a - Be^{-ct}$$

so

$$Z = \frac{1}{c}(a - Be^{-ct})$$

Finally, remember to substitute back for $Z = \ln P$.

$$\ln P = \frac{1}{c}(a - Be^{-ct})$$

It is perhaps neater to leave the solution like this; but it can be written

$$P = \exp\left[\frac{1}{c}(a - Be^{-ct})\right]$$

SAQ 17

If the population were at its maximum level, then growth should be zero. From equation (13), the instantaneous fractional growth rate at population P is

$$\frac{1}{P}\frac{dP}{dt} = a - c\ln P$$

Putting $P = M$ and equating the fractional growth rate to zero

$$a - c\ln M = 0$$

Hence

$$\ln M = \frac{a}{c} \qquad \text{or} \qquad M = \exp\frac{a}{c}$$

The same result can be arrived at equally well by looking at the solution of equation (13) found in SAQ 16 at t tends to infinity.

SAQ 18

From equation (13), the slope $S = dP/dt$ of the solution curve is given by

$$S = P(a - c\ln P)$$

To find a maximum, I first differentiate this equation and find values of P for which $dS/dP = 0$. Using the product rule

$$\frac{dS}{dP} = a - c\ln P + P\left(-\frac{c}{P}\right)$$
$$= a - c - c\ln P$$

This is zero if

$$a - c = c\ln P$$

$$\ln P = \frac{a}{c} - 1$$

$$P = \exp\frac{a}{c} \times \exp(-1) \qquad (23)$$

Using the result of SAQ 17, the slope is zero when

$$P = \frac{M}{e}$$

since the maximum population, M, is equal to $\exp(a/c)$.

I should check that this stationary point I have found is a local *maximum*. I can readily find the second derivative here, so I shall use that test. I get

$$\frac{d^2 S}{dP^2} = -\frac{c}{P}$$

So at the stationary point at $P = M/e$ the second derivative is *negative*, so I have a maximum.

To find the maximum value of the slope I must substitute the value for P I have found into my original equation for the slope

$$S = P(a - c\ln P)$$

My maximum occurs when $P = M/e$, or, going back to equation (23), when

$$\ln P = \frac{a}{c} - 1$$

So the maximum slope is

$$S = \frac{M}{e}\left[a - c\left(\frac{a}{c} - 1\right)\right]$$
$$= \frac{Mc}{e}$$

SAQ 19

Without predation the hare population, H, grows logistically, according to the equation

$$\frac{dH}{dt} = aH\left(1 - \frac{H}{M}\right)$$

In equilibrium, the actual rate of increase of the hare population must be zero, so at any hare population H_0 at which equilibrium can exist, I must have

$$\text{number of hares eaten} = aH_0\left(1 - \frac{H_0}{M}\right)$$

At equilibrium, I have supposed that

$$L_0 \propto H_0$$

49

I have also supposed that in equilibrium

number of hares eaten $\propto L_0$

Therefore, the number of hares eaten is proportional to H_0 and I can write

number of hares eaten $= rH_0$

for some constant r, so that

$$rH_0 = aH_0\left(1 - \frac{H_0}{M}\right)$$

Therefore

$$H_0 = 0 \quad \text{or} \quad r = a\left(1 - \frac{H_0}{M}\right)$$

Hence

$$H_0 = 0 \quad \text{or} \quad H_0 = M\left(1 - \frac{r}{a}\right).$$

On these suppositions I can conclude that the hare population can only be in equilibrium at one level: that is, $H_0 = M(1 - r/a)$. The rather unexciting possibility of equilibrium when $H_0 = 0 = L_0$ can be ignored. Since lynxes were supposed proportional to hares at equilibrium, the lynx population also has just one possible equilibrium level, corresponding to this.

SAQ 20

I can write equation (18) in the form

$$\frac{d^2l}{dt^2} + abl = 0$$

Knowing that ab is positive, I hope you recognize this as the simple harmonic motion equation given in Unit 15. The important point is that its solutions are *oscillatory*.

The solution of the equation is

$$l = A\sin(\omega t + \alpha)$$

where A and α are the arbitrary constants and $\omega = \sqrt{(ab)}$. There is nothing wrong if you wrote your solution in the alternative form $C\sin\omega t + B\cos\omega t$.

I can find h from equation (17), as

$$h = \frac{1}{b}\frac{dl}{dt}$$

$$= \frac{1}{b}A\omega\cos(\omega t + \alpha)$$

SAQ 21

(a) The formula is

$$\sin(\alpha + \beta) = \sin\alpha\cos\beta + \sin\beta\cos\alpha$$

Therefore

$$\sin(\omega t + \pi/2) = \sin\omega t\cos\pi/2 + \cos\omega t\sin\pi/2$$

$$= \cos\omega t$$

So $\cos\omega t$ and $\sin\omega t$ differ in phase by $\pi/2$.

(b) The prediction of the model, in equations (19), is that the oscillations are $\pi/2$ *out* of phase.

(c) So far as one can tell from Figures 12 and 14, there is no difference in phase in the actual oscillations of lynx and hare. So this prediction of the model does not correspond with reality.

SAQ 22

I want to eliminate h from the equations

$$\frac{dh}{dt} = -al \tag{16}$$

and

$$\frac{dl}{dt} = bh + cl \tag{20}$$

I will first differentiate equation (20).

$$\frac{d^2l}{dt^2} = b\frac{dh}{dt} + c\frac{dl}{dt}$$

I can now use equation (16) to replace dh/dt.

$$\frac{d^2l}{dt^2} = b(-al) + c\frac{dl}{dt}$$

This equation can be rearranged as

$$\frac{d^2l}{dt} - c\frac{dl}{dt} + abl = 0$$

SAQ 23

The solution is

$$l = \exp\left(\frac{c}{2}t\right)(A\sin\omega t + B\cos\omega t)$$

where $\omega = \sqrt{(4ab - c^2)}/2$, and A and B are the arbitrary constants. Alternatively, this solution can be written as

$$l = \exp\left(\frac{c}{2}t\right)[C\sin(\omega t + \alpha)]$$

where C and α are the arbitrary constants.

SAQ 24

I have

$$l = A\exp rt$$

and

$$\frac{dh}{dt} = -al$$

Hence

$$\frac{dh}{dt} = -aA\exp rt$$

I can solve this equation by direct integration and the indefinite integral required is a standard integral. I obtain

$$h = -\frac{aA}{r}\exp rt + K$$

where K is the constant of integration.

The value of K depends on the initial conditions: the value of h at $t = 0$. However, the important feature of the behaviour of the solution is independent of K. I know that a, A and r are all positive, so $-(aA\exp rt)/r$ is a *negative* multiple of $\exp rt$—which is a positive exponential. A sketch graph of this solution will look like Figure 18(b).

SAQ 25

To find h, I shall first make it the subject of the equation (20).

$$h = \frac{1}{b}\left(\frac{dl}{dt} - cl\right)$$

To find dl/dt, I must differentiate the equation

$$l = C\exp\left(\frac{c}{2}t\right)\sin(\omega t + \alpha)$$

Using the rule for differentiating a product

$$\frac{dl}{dt} = C\frac{c}{2}\exp\left(\frac{c}{2}t\right)\sin(\omega t + \alpha) + C\omega\exp\left(\frac{c}{2}t\right)\cos(\omega t + \alpha)$$

$$= C\exp\left(\frac{c}{2}t\right)\left[\frac{c}{2}\sin(\omega t + \alpha) + \omega\cos(\omega t + \alpha)\right]$$

Therefore

$$\frac{dl}{dt} - cl = C\exp\left(\frac{c}{2}t\right)\left[-\frac{c}{2}\sin(\omega t + \alpha) + \omega\cos(\omega t + \alpha)\right]$$

and the predicted hare population is

$$h = \frac{1}{b}\left(\frac{dl}{dt} - cl\right)$$

$$= \frac{C}{b} \exp\left(\frac{c}{2}t\right)\left[-\frac{c}{2}\sin(\omega t + \alpha) + \omega \cos(\omega t + \alpha)\right]$$

Although c is negative in the case of interest, so that there is a positive multiple of $\sin(\omega t + \alpha)$ in this solution, the oscillations of hare and lynx are still *out* of phase, because of the presence of the cosine term.

SAQ 26

The logistic equation is based on the supposition that there is a fixed maximum, or equilibrium, population M that the environment can support. This, by implication, supposes that the environment is constant; that available food is constant, and so are any other factors vital to the species' survival. Random variations in these factors from year to year were deliberately ignored. The important point here is that the possibility of the species affecting its environment is overlooked. A population might increase above its equilibrium level, but at the expense of over-using its food source. Thus M would be temporarily decreased. Eventually, the population would decrease as food becomes scarce, then later the food species would recover, and M increase again, and so on. This is just the sort of possibility—leading to oscillatory population levels—that I have been discussing.

Incidentally, I would not claim that the above is the only 'right' answer to this question. More succinctly you might argue that the model in Section 3.3 was set up on the supposition that there was a *maximum* population that the environment can support. By implication the population is supposed never to go above that level. Had the model been set up on the supposition that there is an *equilibrium* population that the environment can consistently support, then the possibility of population variations such as those shown in Figures 13(b) and (c) in Population in *Modelling Themes* might not have been overlooked.

SAQ 27

(a) There is no 'right' answer to this equation. Various models are reasonable. I shall set up just one of them. First note the following points about the data.

  (i)   Both graphs are 'oscillatory'.

  (ii) They are in phase.

  (iii) The graph of the prey population has a *logarithmic* scale, but that of the skua population is linear.

These lead me to set of a model on the following suppositions. I shall write S for the skua population and R for the *logarithm* of the rodent population. I shall suppose that the number of skua births is proportional to R; the number of skua deaths is proportional to S; and the oscillations of R are sinusoidal.

The second supposition seems particularly reasonable here in the light of the information given. The skuas' response to a rodent shortage at the breeding grounds is not to stay there and starve, but to return to sea. So births decrease, but deaths do not increase.

On these suppositions I form the model

$$\frac{dS}{dt} = \text{births} - \text{deaths}$$

$$= \alpha R - \beta S$$

where α and β are constants

If the oscillations of R are sinusoidal, I can write

$$\frac{dS}{dt} = \alpha(R_0 + A\sin\omega t) - \beta S$$

where R_0 is the average level of R; A the amplitude and ω the frequency of the prey oscillations.

(b) The solution to the model I have set up in (a) is oscillatory. It is in fact

$$S = L + M\sin(\omega t + \varepsilon) + N\exp(-\beta t)$$

where $L = \dfrac{\alpha R_0}{\beta}$; $M = \alpha A\big/\sqrt{(\omega^2 + \beta^2)}$; $\tan\varepsilon = \dfrac{-\omega}{\beta}$;

N is an arbitrary constant. (I do not suggest that you check this!)

For large values of t the term $N\exp(-\beta t)$ is small and the oscillatory term dominates.

(c) The birds vary their birth rate in response to the available food supply, which is how I defined 'natural control'. The result of the skua's 'natural control' behaviour is to make the extent of the oscillations in its numbers much less than the extent of the oscillations in those of its prey. However, I do not have time to show you this!.

Modelling by Mathematics

ACKNOWLEDGEMENTS

Grateful acknowledgement is made to the following for material used in this Unit:

Tables

Table 1 from D. Lack, *Population Studies of British Birds*, Oxford University Press 1966; *Table 4* from R. Hudson, 'The spread of the collared dove in Britain and Ireland' in *British Birds*, Vol. 58 1965 Macmillan.

Figures

Figures 1(b), *3(b)*, *4*, *5*, *8(a)*, *12 and 13* from M. Williamson, *The Analysis of Biological Populations*, Edward Arnold 1972; *Figure 8(b)* from A. J. Lotka, *Elements of Mathematical Biology*, Dover 1956; *Figure 14* from M. Edel, *Economics and the Environment*, Prentice-Hall Inc. 1973; *Figure 20* from M. Andersson, 'Population ecology of the long-tailed skua' in *Journal of Animal Ecology*, Blackwell 1976.